From Mulberry Leaves to Silk Scrolls

The Lawrence J. Schoenberg Studies in Manuscript Culture

VOLUME I

Series Editors

From Mulberry Leaves to Silk Scrolls

New Approaches to the Study of Asian Manuscript Traditions

EDITED BY

Justin Thomas McDaniel and Lynn Ransom

The Schoenberg Institute for Manuscript Studies
UNIVERSITY *of* PENNSYLVANIA LIBRARIES

Distributed by University of Pennsylvania Press | *Philadelphia*

Distributed by
University of Pennsylvania Press
Philadelphia, Pennsylvania 19104-4112
www.upenn.edu/pennpress

Book design and composition by Judith Stagnitto Abbate / Abbate Design

Printed in the United States of America on acid-free paper
10 9 8 7 6 5 4 3 2 1

Library of Congress Cataloging-in-Publication Data
From mulberry leaves to silk scrolls : new approaches to the study of Asian manuscript traditions / edited by Justin Thomas McDaniel and Lynn Ransom.
pages cm. — (The Lawrence J. Schoenberg studies in manuscript culture ; volume 1)
"Several of the essays in this volume were first presented at the Fourth Annual Lawrence J. Schoenberg Symposium on Manuscript Studies in the Digital Age, 'Writing the East: History and New Technologies in the Study of Asian Manuscript Traditions,' which was held October 21-22, 2011, at the University of Pennsylvania and the Free Library of Philadelphia"—Preface.
Includes bibliographical references and index.
ISBN 978-0-8122-4736-7 (alk. paper)
1. Manuscripts—Asia. 2. Manuscripts, Oriental. 3. Codicology. I. McDaniel, Justin, editor. II. Ransom, Lynn, editor. III. Series: Lawrence J. Schoenberg studies in manuscript culture ; v. 1.
Z106.5.A78F76 2016
091—dc23

2015007467

IN MEMORIAM

Lawrence J. Schoenberg

July 1, 1932–May 7, 2014

Contents

Part II. Inscribing Religious Practice and Belief

Part III. Technologies of Writing

Preface

THIS COLLECTION OF ESSAYS inaugurates a new series for the field of manuscript studies: the Lawrence J. Schoenberg Studies in Manuscript Culture. The Schoenberg name has a long history of use at the University of Pennsylvania Libraries due to the generosity and vision of Larry Schoenberg (C'53, WG'57), who sadly passed away in 2014 before seeing the first volume of the series published. The impact of Larry and his wife, Barbara Brizdle, on manuscript studies at Penn has been felt through the establishment of the Schoenberg Center for Electronic Text and Image in 1996 and the creation of the Schoenberg Initiative in 2006 to assist the Libraries in purchasing new manuscripts, and the annual Schoenberg Symposium on Manuscript Studies in the Digital Age, which began in 2008. In 2011, Larry and Barbara donated their collection of manuscripts to the University of Pennsylvania Libraries, which led to the founding of the Schoenberg Institute for Manuscript Studies in 2012. Consistent with Larry's vision of sharing his collection and the knowledge gained through studying manuscripts, this new series will bring together scholars from around the world and across disciplines to present research related to the study of premodern manuscripts and to consider the role of digital technologies in advancing manuscript research. Whether relying on traditional methods of scholarship or exploring the potential of new technologies, the research presented in each volume will highlight the value of the manuscript book in understanding our intellectual heritage.

Several of the essays in this volume were first presented at the Fourth Annual Lawrence J. Schoenberg Symposium on Manuscript Studies in the Digital Age, "Writing the East: History and New Technologies in the Study of Asian Manuscript Traditions," which was held October 21–22, 2011, at the University of Pennsylvania and the Free Library of Philadelphia. The symposium covered a range of issues relating to Asian reading and writing cultures, especially as they pertain to the manuscript source. The success of the event inspired the editors to invite contributions from other scholars. The resulting collection of essays explores such topics as best practices for preservation and cataloging; demonstrates the value of collaboration among scholars who work on different aspects of codicological, paleographic, orthographic, and material culture studies; and reveals how these material objects were used for religious, political, cultural, and pedagogical purposes. Whereas manuscript studies in the West have benefited from a long history of scholarship, scholars of Asian manuscript traditions have only recently begun to excavate this rich field of study. As their work continues, their research can only enhance our understanding of manuscript culture. It is fitting, then, that the first volume of the Lawrence J. Schoenberg Studies in Manuscript Culture begins its work in the area of Asian manuscripts by giving scholars the opportunity to share their work and advance our knowledge.

We also acknowledge here our gratitude to those who made the publication of this volume possible. Our first thanks go to the volume's co-editor, Justin McDaniel, Professor of Religious Studies at the University of Pennsylvania, who first approached us with the idea to devote an entire symposium to Asian manuscript traditions. We would also like to thank H. Carton Rogers, Vice Provost and Director of Libraries at the University of Pennsylvania, and Jerome Singerman, Senior Humanities Editor at the University of Pennsylvania Press, who through their generosity and good will make possible the continued publication of volumes in this series. In gratitude for everything that the Schoenbergs have done, we dedicate this volume to the memory of Larry Schoenberg and to Barbara Brizdle, whose ongoing support ensures the continuation of the good work that Larry began. We can repay his generosity only by spreading his vision as widely as possible. We offer this series as a small contribution toward that enormous debt.

Lynn Ransom
Schoenberg Institute for Manuscript Studies
University of Pennsylvania Libraries

Introduction

Justin Thomas McDaniel

IN THE READING ROOM of the Chester Beatty Library in Dublin, in 2012, I unwrapped an old mulberry-paper manuscript that the library's catalogue described as a Phra Malai text from Siam written in thin *mul* script. As I gently paged through the concertina, or folding-style, manuscript, marked by water stains, worn edges, and fading illuminations, I discovered that not only was the catalogue description incorrect; this was not even one text. The manuscript contained illustrations of seemingly drunken Buddhist monks chanting while children played dice-like games in front of them, amorous couples stealing embraces in monasteries, hunters shooting animals, women smoking opium, a man stirring a cauldron full of human heads, golden reliquaries surrounded by colorful flags, monks sweeping a cloister, heavenly beings, a Chinese opera performance in front of a crowd of boisterous children, lecherous men, fighting thugs, and terribly neglectful parents. A series of scenes depicted half-naked women seducing a man in a forest and a tree from which naked women hung as fruit. I even had a postmodern moment when I turned a page and saw a drawing of a monk reading a manuscript while another monk transcribes it. There were other paintings as well—woodland creatures, birds, flowers, and mountains. I was perplexed; based on the catalogue title and library record, this was not the manuscript I expected.

I rubbed my eyes, shook my head, and dove back in. One side of the manuscript was indeed a version of the Phra Malai text, but only one painting seemed to be even remotely connected to that story. The other side of the manuscript was a completely different text: the Matika of the Dhammasaṅganī of the Abhidhamma. The Phra Malai is in vernacular Thai, while the Matika is in the Pali language. There was evidence that at least two, if not three different scribes worked on the text—and perhaps as many as four different artists. However, only one author, Phra Dhammavuḍḍhi Bhikkhu, was named (in a scribal hand different from that of the text), and no date or patron was listed. Some paintings were left half-finished. The text and images did not match, and I could not discern any sensible order to the paintings. I tried to understand how these random images and two texts might have gone together. The scribes who worked on the text were well trained; the different scribal hands were all steady; spelling and grammar had normal inconsistencies found in most manuscripts but nothing strikingly awful. These were not amateurs. The illuminations were of mixed quality but not particularly sloppy. One of the painters had an unusual talent for adding detailed patterns to silk sarongs and flower petals. The manuscript was made of high-quality and thick mulberry (*khoi*) pulp, and the wide variety of pigments used revealed a patron with disposable income. Perhaps some of the scenes, like the fighting thugs, drunken (or perhaps ignorant) monks, lecherous men, gambling, and amorous couples, were supposed to represent different levels of hell.

The story of Phra Malai recalls Dante's journey past the abyss and into various levels of hell: he descends through the levels and sees the *contrapasso* punishments for those who are greedy, hateful, deceitful, violent, and filled with lust. The story has been popular with storytellers, muralists, scribes, and illuminators for centuries in Thailand. Nearly universally, the hells are depicted in manuscripts very graphically, with severed limbs, burning flesh, torture chambers, and voracious animals and vegetation. Precedents are lacking, however, for scenes of realistic vices, gambling, sex, addiction, or fighting as metaphors for levels of hell. Perhaps I was grasping at straws. Why would illuminations of heavenly scenes and peaceful nature scenes be intermingled with depictions of misbehaving monks and laypeople? Moreover, the Matika, though it is a popular theme for scribes, has nothing to do with the story of Phra Malai. It is a Pali liturgical text that is chanted at funerals. The Abhidhamma text and its commentaries, however, lack specific instructions on how to conduct rituals, nor does their content have any relation to rituals. The Phra Malai is a vernacular Thai text about levels of hell; the Pali Abhidhamma Matika is about the nature of the mind: the content and language of the texts are quite different, although both are often recited at funerals and cremations.

Looking for more clues, I returned to the manuscript; I found one on the first leaf. Attached to the inside of the cover was a little label in English, typed on a very old typewriter. It read simply, "Siamese manuscript written in Cambodgian character. On the occupations, sports, agriculture etc. of the Siamese. With 15 double leaves illustrated. 22.10 pounds." The original catalog number, 2456, had been crossed out by a later collector; a shelfmark, Ms. 1329, was amended to read Ms. 1330. The inside cover (which is often blank in Thai manuscripts in general) is inscribed "Phra Dhammavuḍḍhi Bhikkhu," followed by the Thai verb *khian* (to write) in red ink and in a hand different from those in the main text. On the basis of that evidence, I have come to believe, although I am still unsure, that this manuscript was actually made for a Western collector. The label clearly reflects a person who could not read the text's content. This is not a text about sports or agriculture. Some of the scenes, hunting and game playing, could be interpreted as a guide to local sports, which might account for the description in the English label; however, the textual content and the vast majority of the illustrations have nothing to do with sports or agriculture. This would have been an unusual acquisition because it is different from most of the Phra Malai, Abhidhamma, Jātaka, or Kammavācā manuscripts available in the markets of Southeast Asia at that time (and today). The fact that all of these different scenes are contained in one manuscript makes it very rare. Did a monk or an artist take a partially completed Phra Malai manuscript and adapt it for sale to a foreign collector? Was this designed as a contrived cultural product rather than for ritual use?

Half-written or partially illuminated Thai manuscripts appear frequently in collections. Large manuscripts often start off beautifully but lapse into incomprehensibility and sloppy execution after the first few leaves. Some manuscripts were clearly too inconsistent and amateurish to be presented as gifts or used in rituals; perhaps they became "practice" manuscripts through which artists and scribes could hone their craft. This would explain a text that was executed by several different scribes. The two texts that compose the Chester Beatty manuscript, Phra Malai and Matika, are relatively consistent in grammar and spelling. A teacher might have read the texts aloud while apprentice scribes transcribed the recitation. Thai manuscript illuminators and scribes rarely worked together. The illuminations were probably intended for the vast majority of Thai monks, nuns, and laypeople who were unable to read Khmer, Khom, Mul, or "Cambodgian" script or who were unfamiliar with Pali, whereas the text was intended for a monk or nun familiar with the Khom script and both vernacular and Pali languages. The paintings were probably executed later (scribes in Thailand always left room for paintings in the large margins of the manuscript) as assignments given by a teacher to a student; or the salacious and entertaining scenes might have been painted in order to fetch a good price from a collector. The fact that several of the paintings are unfinished (as attested by the pencil lines and partially painted figures) suggests either possibility. Many Western collectors acquired manuscripts in Bangkok during the late nineteenth and early twentieth centuries; the manuscript's style dates it to the 1880s. Or the students or students who were working on the manuscript paintings for practice might have been called to other duties and left their work unfinished. (The lives of monks and artists in cities are often uncertain: they are relocated to different monasteries by their abbots with little warning; they have family crises and need to disrobe or return home quickly; they fall in love and lose interest in their duties or training.)

This Thai manuscript composed in Pali and Thai, written in the Cambodian script, and found in an Irish library by an American translator held many possibilities indeed, and it illustrates some of the questions that arise when a scholar unwraps a manuscript. The nature of the medium, the writing tools, the available materials, and the professionals that work on the texts all contribute to often cacophonous cultural products. This manuscript's many voices and many sources pull the text apart. It gives collectors, catalogers, conservators, translators, and art historians the gift of very long days spent on very few pages. Not only does this one manuscript contain many different voices; they are also largely faceless. We have only one name, and this name could have been added later. The identities of the scribes and the illustrators are unknown, as are those of the patron, the producer of the paper, the mixer of the pigments, and the planter of the mulberry tree. We do not know how many hands the text passed through before being purchased by Chester Beatty's in situ collector. We have one box and one label, both added later. The number of anonymous hands that handled the text will most likely never be known. However, it is these unknowns and the skills needed and years spent trying to find answers that make the study of manuscripts fascinating. By sourcing the materials, comparing the orthography, translating the inconsistent texts, and tracing the chain of collectors, we can actually answer much larger questions about local economics and divisions of labor, trade and material history, cultural preferences, the rise of "Asian" collections in the West, and religious requirements. Manuscripts are much more than the semantic meaning of the words they contain. The intellectual reward is in the struggle to figure out the answers one at a time.

The scholars who have contributed to this book perfectly illustrate this struggle. Each essay reveals as much about the questions raised by the study of any one manuscript as the answers their pages provide. Each essay explores premodern handwritten texts for information about the intellectual and cultural contexts from which they emerged, as well as how those contexts changed over time as the manuscript moved from hand to hand, monastery to monastery, and library to library.

The first section of this collection is called "The Art of the Book." Its three chapters focus on the manuscript cultures of Southeast Asia to reveal the variety of ways of thinking about manuscripts as *objets d'art.* This hopes to be a corrective to past studies of manuscripts in this region, which have primarily mined the texts for historical content, ethical treatises, legal and religious doctrines, and courtly literature. The manuscripts examined also expand our understanding of what constitutes the religious and literary curricula and canons of the region. The primary content of the monastic education and religious activity for the various literary, court, and monastic communities in Burma (Myanmar), Cambodia, Laos, Thailand, and the Tai-speaking region of Southern Yunnan (China) consisted of ritual (both protective and daily monastic liturgies), grammatica, legal texts, medical and astrological guides, local histories, and ethical and romantic narratives. The collections are dominated by vernacular and Pali narratives from canonical and noncanonical Jātakas and Dhammapada-aṭṭhakathā anthologies, vernacular stories like the Madhurāsajambū, ritual texts such as the Sattaparitta (Sutmon) and the Kammavācā-uppasampadā, and grammatical texts. In some collections historical and cosmological texts were numerous. The choice of texts can tell us a great deal about the needs and values of the communities that composed or used them. The choices reflect those aspects of religion and culture that were deemed more important and most necessary to teach, especially in the vernacular. These texts may be seen as articulations of a highly local understanding in an attempt to make sense of a world in which Buddhism constituted the overarching, dominant system of ideology and practice; such texts negotiate between the classical and the vernacular, the translocal and the local. They incorporate local elements into a nonnative literary structure. Further research along these lines might lead us far in determining the different modes of interaction between these shifting epistemes. The three contributors to this section offer stunning images and in-depth analysis of individual illuminated and ornamented manuscripts in order to highlight larger trends in the field.

Hiram Woodward, the doyen of Southeast Asian art history, offers a close study of the elephant manuscript tradition in Thailand. While all cultures in Southeast Asia prize the elephant as a symbol of royalty, power, and beauty, in Siam (Thailand before 1939), the lauding of the elephant became both an art and a science. Woodward looks at several examples of the Thai genre called the Tamrā Chāng (or "Characteristics of Elephants"). Woodward was the curator of the impressive Asian collection at the Walters Art Museum in Baltimore for most of his career and draws on the museum's fine Thai manuscript collection for many of his examples. He shows that by looking at one particular type of manuscript tradition unique to Thailand, in his case, the Tamrā Chāng it is possible to speculate broadly on the way manuscripts were produced by the royal court. The text belongs to a particular genre, found also in India (there are treatises on the characteristics of horses and cats as well), and the illustrations exemplify the highly refined court taste of the first half of the nineteenth century. Therefore, this study is both narrow and broad. What I particularly was struck by in Woodward's highly detailed and clearly written study is that the authors of these elephant manuscripts did not distinguish between divine or mythological elephants and actual examples of different elephants seen in nature. The blurring of lines between the worlds of literature, religion, and science are witnessed in this genre of texts.

Alexandra Green also looks at the way literature and history, folklore and science, are mixed in the manuscript cultures of Southeast Asia. Known earthly realms replete with structures, vegetation, and animals seen in Burma abut detailed scenes of hell realms, paradises, and sacred mountains. These scenes are accompanied, unlike in the manuscript traditions of Siam, Laos, and Cambodia, with detailed textual description of these scenes. Green not only examines the illuminations and the text of a hitherto unpublished Burmese cosmological manuscript in the British Museum, but also compares this manuscript with representations of the cosmos on monastic murals and architectural ornament. Although she does not discuss this issue explicitly, her study forces the expert in manuscript art to think about not only the ways in which the medium (size, shape, tensile strength, pigment, ink, and the like) changes the possibilities of representation in manuscripts but how temple walls, ceilings, and architraves create other possibilities. She argues that representations of the cosmos describe ideas about how the universe functions; the manuscript she discusses not only reflects religious and social beliefs (in particular, the law of cause and effect) but also would have generated merit to the manuscript's donors, contributing to the goal of nirvana. These manuscripts therefore do not work simply to amaze and delight the reader with the variety and complexity of the universe, but actually ground her or him in the comfort of the karmic system.

Sinéad Ward also works with Burmese manuscripts but from a very different perspective and on a very different genre. She looks at the narrative scenes in the decorative Kammavācā manuscripts. The Kammavācā genre is one of the most common genres of manuscripts in the region, describing the basic procedures for conducting the major monastic rituals: ordination, robe offerings, liturgies, and the like. However, Ward has found a manuscript in the Asian Art Museum in San Francisco that contains not only the text of the Kammavācā but also illuminated narrative scenes. Like the manuscript I mentioned at the beginning of this introduction, the textual content and the illumination have nothing to do with each other; most likely the artist and the scribe did not work together. Unlike in Siamese manuscript traditions, however, these narrative scenes are rare (appearing in only 10 of the 440 manuscripts surveyed) in the Burmese tradition; their introduction in the Kammavācā marks a departure from the ornamental tradition. Kammavācā manuscripts were actively collected by foreigners (particularly British colonial officials in the late nineteenth century and later by international art and rare-book collectors) because of their beautiful ornamentation. These narrative scenes can be seen on the one hand as an extension of ornamentation, but also as pedagogical on the other. Ward offers a clear introduction to this widely seen but rarely studied genre, and offers some suggestions about ways of thinking about the rare examples of narrative on these otherwise straightforward monastic procedural texts.

In the second section, "Inscribing Religious Practice and Belief," we turn away from a focus on the art of the manuscripts to look at the ways in which manuscripts were used in Asia to enhance religious belief and guide religious ritual. We also change our regional focus and move north to northern Thailand and China. Just as the studies in the first section broke down the divisions between religious and secular subjects, as well as canonical and extracanonical works, the studies in this section break down the divisions between religious traditions, namely Taoism, Buddhism, animism, and Confucian schools of thought and practice. They also reveal the importance of new discoveries of manuscript caches in forgotten archives, caves, and monasteries.

Angela Chiu has been examining a large body of vernacular manuscripts in northern Thailand that discusses local history and that has been largely overlooked in favor of Pali texts. She also approaches the study of Southeast Asian manuscripts from a very different place and perspective than the contributors to the first section of this book. Instead of focusing on the material and illustration of her manuscripts, she focuses on how the historical content of the text exposes the ways in which authors in the region used text to establish institutions and lineages. She works on the historical tradition of northern Thailand, Laos, and the Shan region, showing that chronicles from this region were not merely repositories of royal lineages, battles, property claims, and place names, but were agents in the construction of geographical and ritual space. Many of the genre of manuscripts called *tamnan,* or chronicle, on which she focuses are about the origins of Buddhist relics in the region. Twelve famous monasteries in northern Thailand claim to have relics of the physical body of the historical Buddha; other monastery chronicles often make the claim that the Buddha visited (often by magically flying) particular monasteries in the past. Chiu looks at largely understudied manuscripts like *The Chronicle of the Buddha Image Lying on a Mango-tree Log*, *The Chronicle of Wat Suan Dok*, and *The Relic History of Wat Haripunchai*, among others, to demonstrate how these noncanonical chronicles of the Buddha's visits to local sites in northern Thailand reflect a mechanism through which Buddhist values were integrated with their own values by the La Na people.

Ori Tavor, a specialist in early Chinese Taoist and Buddhist ritual, as well as the Confucian principles that inform much early writing on ritual in China, focuses on manuscripts that place the body of the ritualist front and center. He is one of the first scholars to look intensely at a group of 1,200 bamboo-strip manuscripts from the Warring States period (453–211 BCE) that were recently discovered in the Hong Kong antiquities market and are now held at the Shanghai Museum. This large group of manuscripts has been neglected by many because they are difficult to date and offer little historical information. However, Tavor shows that while they are of arguable benefit to historians, they enable scholars of religion and intellectual history to reconstruct the development of two competing modes of religiosity: a practical religiosity associated with a give-and-take approach to ritual and a moral religiosity that reflected a theology based on devotion to ethico-religious guidelines. Tavor argues that a reading of these manuscripts against the background of received sources shows that these two modes ultimately coexisted through the Han Dynasty and were instrumental in the development of organized religious traditions.

Daniel Sou's essay is a perfect complement to Tavor's work because Sou shows that the give-and-take approach and the pragmatic theology derided by the elite's focus on the body were still popular religion during the Han Dynasty. In order to make that argument, he also relies on new manuscript evidence. Instead of focusing on the elite, though, his manuscripts, which were found in a series of newly excavated tombs and known as the Shuihudi, Baoshan, and Guanju bamboo manuscripts, reveal the religious lives of the common people. Instead of the sophisticated intellectual argument about the connection of the body of the ritualist to the structure of the universe, Sou's sources, especially the manuscript of the Book of Days, deal with the practical concerns of hunger and disease. His detailed examination of the practice of exorcism shows how larger theories about the relation between the body and the unseen worlds were expressed in ritual and explores the political purposes that such texts might have served in maintaining social order and controlling local customs.

From the broad implications seen in Chiu, Tavor, and Sou, the third section of this collection, "Technologies of Writing," returns to the details of the writing technologies themselves and ends with two essays that consider the impact of new technologies in the study of manuscripts. In the first place, the material foundations of manuscripts (palm leaf, birch bark, vellum, papyrus, mulberry, skin, wood-pulp paper), as well as writing and drawing tools (pigments, bindings, metal inscribing tools, wooden reading pointers, and the like) have agency. Palm leaf resists the absorption of ink but is an ideal support for inscribing with sharp styluses. Mulberry paper can be dyed, folded, and stacked, but it absorbs moisture too easily and is susceptible to mold. Vellum is expensive to produce in large quantities. The properties of silk and rice paper make corrections difficult. Each material text has lineages and ideal types of art that determine, consciously or unconsciously, the possibilities of creation: namely, how writers or artists are governed by the power of previous exemplars in their tradition. The technology of writing can change very slowly; and this is a problem when new types of knowledge demand new types of textual formats and materials.

Kim Plofker looks at the problem of using traditional Indian Sanskrit manuscript technology, which was originally used to record orally and aurally produced texts as the vehicle for scientific texts. These scientific texts required drawings, charts, chains of numerals, and the like, which were not suited to a manuscript culture that replicated oral verse literature through long strings of uninterrupted syllables. Drawing on evidence from both Al-Bīrūnī's observations of Indian manuscript technologies and work on a Sanskrit Purāṇa text in the collection of the University of Pennsylvania, she traces the development of the bhūta-saṅkhyā, or "object number" system, back to the difficulty of fitting "number words" into a metrical verse structure, and shows how these difficulties were circumvented by an ingenious (albeit impractical) system of numerical equivalents. Her contribution is an excellent example of seeing how both the technology of the actual material of manuscripts and the weight of a literary tradition affected the very development and preservation of scientific knowledge.

Sergei Tourkin approaches the technology of writing from a different perspective. He shows that scholars are also controlled by prototypes, in his case, the system of abbreviation in Arabic and Persian manuscripts. While the abbreviations of Arabic and Persian words used in manuscripts have long been known by scholars, they might assume the same abbreviations are used by all writers across the same tradition. That this is not the case he shows by looking at four astrological and astronomical manuscripts (primarily in the almanac and personal horoscope genres), in Persian manuscript culture. Different writers working with zodiac calculations used different abbreviations for the same word. In some systems the last two letters were used, while in others only the last letter, and in others the last radical consonants of the roots. Besides highlighting the importance of not depending on tradition to assume how authors used abbreviation, Tourkin shows that the physical agency of the manuscripts themselves—the height of the rows or the width of the columns—often determined what abbreviations were chosen. The importance of this discovery is clear when needing to read exact dates and numbers in these manuscripts and could have only been determined by understanding the material agency and writing technology of the texts themselves.

With Chapter 9, we turn to the role of new technologies in manuscript studies. The importance of a digital technology is seen in Susan Whitfield's overview of the International Dunhuang Project (IDP). This project is the gold standard in the field of Asian manuscript studies. She shows that starting in 1998, the IDP has placed over 100,000 manuscripts online. These manuscripts, taken from the more than 150,000 manuscripts spread over twenty different international manuscript collections, draw from more than twenty different languages and scripts, such as Sogdian, Syriac, Uygur, Tibetan, Judaeo-Persian, Greek, Chinese, and many others, composed and read by intellectuals from the Manichaean, Buddhist, Taoist, Nestorian Christian, and Zoroastrian religious traditions. The manuscripts were found in a mixture of caves, personal collections, monasteries, and even rubbish heaps. They cover nearly every conceivable subject—law, religion, commerce, astrology, medicine—and use wood, silk, birch bark, palm leaf, paper, and other material supports. This wide variety of subjects, languages, and texts requires teams of scholars, and in order to access and discuss these manuscripts, technology (imaging, coding, cataloging, carbon dating, material analysis) is absolutely essential. Complementing the highly detailed examination of texts by scholars like Ward, Woodward, Plofker, and Tourkin, Whitfield's summary of the IDP shows the value not just of laborious and often lonely scholarly work on individual manuscripts, but the highly accessible and collaborative work that is possible with large projects. Manuscripts can be seen in much larger comparative contexts where new questions are raised and new technologies are employed.

The section ends with Peter Scharf's essay "Providing Access, Promoting Study, and the Impact of the Digital Age," which turns to the very practical problem of how one searches through thousands of manuscripts, not only for titles and subjects, but for individual terms, phrases, and even grammatical usages. Scharf's work has long been heralded for helping other scholars do just this. His painstaking work at the Sanskrit Library (sanskritlibrary.org) provides access to hypertexts of Sanskrit manuscripts. He led the development of a "comprehensive integrated hypertext catalogue and software to integrate digital images of manuscript pages with the corresponding machine-readable text, thereby providing direct and focused access to specifically sought passages on individual manuscript pages." Not only are the images of manuscripts provided, but also the metadata to enable easy searches. He undertook this very large project in order to help preserve the manuscripts and make them available outside of specialists' archives. It is work like this that will allow future scholars to compare texts easily and develop best practices for investigations into manuscript cultures in the future. Scharf describes the project in detail and then provides examples of possible searches and uses of the digital catalogue and hypertexts.

Lynn Ransom and I, along with the support of the Schoenberg Center for Manuscript Studies and the Kislak Center for Rare Books, Manuscripts, and Special Collections and its wonderful directors and staff, took on this project to offer a glimpse of the recent innovations into the study of Asian manuscript traditions. We believe it reflects the great advances that have been made in the field over the past decade but also, dauntingly, shows us and the readers how much work there is still to be done.

PART I

The Art of the Book

CHAPTER 1

The Characteristics of Elephants

A Thai Manuscript and Its Context

HIRAM WOODWARD

THE ILLUSTRATED THAI TREATISES on the characteristics of elephants are quite similar to each other. One dated equivalent to 1815 is in the National Library, Bangkok. Others have dates of 1816 (Chester Beatty Library, Dublin) and 1819 (Staatsbibliothek, Berlin). Undated copies include ones in the Harvard Art Museums, thought by Henry Ginsburg to date from the 1790s; the British Library; and the Walters Art Museum, Baltimore.[1] Illustrations in this essay are taken from the Walters manuscript, which was presented to the museum in 2002 by the Doris Duke Charitable Foundation's Southeast Asian Art Collection. All the elephant characteristics manuscripts are accordion pleated, on paper, with writing and illustrations in ink, opaque watercolors, and gold.

Quite aside from the grace and energy of the outlines to be seen in the illustrations, the Walters manuscript and the others like it are of scholarly interest for several reasons: because they epitomize the "characteristics" literature; because more can be said about their social function than is the case with many other Thai manuscripts; because of what they reveal about the relationship of beliefs promulgated by Buddhists to those held by court Brahmans; because they include depictions of presiding deities with intriguing characteristics; and because they open unexamined windows on the religious practices of old Siam.

Characteristics

Although details can vary from text to text, the elephants are considered to constitute a set list. Each of the elephants belongs to a family—Brahma (ten elephants), Vishnu (eight elephants), Iśvara (that is, Shiva, eight elephants), or Agni, the Hindu god of fire (forty-two good and eighty evil elephants). The families correspond to the Indian castes. There are also elephants of the eight directions of space, plus another fourteen that are their offspring.[2] Accompanying each illustration in the manuscripts is a short poem. The verses in the Walters manuscript correspond imperfectly to those that appear in a manuscript dating from 1782, the year Bangkok was established as the capital. This text is entitled *Tamrā laksana chāng kham khlong* (*The Treatise on the Characteristics* [*lakṣaṇa*, a loanword from Sanskrit] *of Elephants* in the *khlong* verse form), and it appeared in a printed edition in 1938.[3] An older text, in a different verse form, dates from 1738.[4] In the Chester Beatty Library manuscript (which is also based upon the 1782 text), tiny numerals preceding the verses indicate the number of each elephant in its class.

Near the beginning of the manuscripts (following the three presiding deities and the three celestial elephants, discussed below) are the ten elephants of the Brahma family, who also represent the ten herds of the Himalaya forest (or *Himaphan*, as it is called in Thai). The very first of these is Chaddanta (Figure 1.1). The accompanying verse says,

> A blessing and a reward, an albino as clear as beautiful silver;
> Spreading glory over the three luminous worlds;
> Essence of good fortune, by name Chaddanta, proclaimed in ages past;
> Mounted by the shining world emperor superior in power.[5]

The name *Chaddanta* figures in Pali-language Buddhist scriptures in several different senses: it is a name of a lake on the slopes of the Himalayan mountains, of a forest there, of a tribe of elephants, and of the king of this elephant tribe. The Buddha, in one of his previous lives, had in addition been born as the elephant king Chaddanta.[6] A hunter was sent to cut off the elephant king's tusks as an act of revenge but was unable to carry out

Figure 1.1. The first two elephants in the world of humans: Chaddanta (left) and Uposatha (right), with accompanying verses. "The Characteristics of Elephants," accordion-pleated manuscript, Bangkok, Siam, ca. 1810–30, face 1, f. 9. Walters Art Museum, Baltimore, gift of the Doris Duke Charitable Foundation's Southeast Asian Art Collection, 2002 (W.893).

the task. Out of compassion, Chaddanta managed to saw off his own tusks to present to the hunter.

In Thai poetry, the emphasis is on euphony and the repetition of sounds. It is nearly impossible in translation to capture both the sense and the quality of the language. The best one can do is to point out poems in English that have similar sound patterns, such as the poems of Gerard Manley Hopkins, which are somewhat comparable in the role of wordplay:

> Left hand, off land, I hear the lark ascend,
> His rash-fresh re-winded new-skeinèd score
> In crisps of curl off wild winch whirl, and pour
> And pelt music, till none's to spill nor spend.
>
> ("The Sea and the Skylark," ll. 5–8)

Most Thai poetry, however, is formulaic and does not have the startling quality of Hopkins's verse.

In the Walters manuscript, the elephants look much alike, and the artist has more or less ignored the different qualities elucidated in the poems. In the Chester Beatty Library manuscript, the evil elephants that appear at the very end are individualized and depicted in lively postures.[7] Since the Walters manuscript was never completed, however, it is hard to say whether repetition would have ruled the day from beginning to end, although the unfinished state does make it possible to observe the artist's (or artists') working methods (Figure 1.2).

There are reports of unpublished Thai manuscripts of a different sort, which contain practical information about elephants. Such manuscripts may resemble the Indian text titled *Mātaṅga-līlā*, which served to some degree as a useful handbook. It nevertheless also contains descriptions of desirable *lakṣaṇa*, characteristics that could be considered the ancestors of

Figure 1.2. An unfinished page. Walters Art Museum W.893, face 2, fol. 20.

the Thai verses: "Whose right tusk tip is high(er than the left), whose mighty trunks and faces are marked with (light) spots, whose stout fore and hind legs have invisible joints, these (elephants), O prince, are fit vehicles for you."[8] But in contrast to the Thai text, the elephants with such traits are not given names.

As Henry Ginsburg pointed out in his book *Thai Manuscript Painting*, the models for these Thai texts were not the old Sanskrit *Mātaṅga-līlā* but more recent Sanskrit and Tamil texts from southern India.[9] These texts also contain depictions of elephants with different traits. A related class of southern Indian texts concerns the characteristics of horses. These have a direct counterpart in Thai horse manuscripts, both in regard to the style of illustration and text. The southern Indian treatises give descriptive names to horses of different appearance, for instance, and so in this regard much resemble the Thai elephant manuscripts:

> *Nāgajātīya*: with circular spots of multiple colors all over the body
>
> *Jyōtiṣmān*: black, but white in chest, face, hooves, mane, eyes, and back
>
> *Pañcabhadraka*: white only in the four hooves and face[10]

The Three Celestial Elephants

The Walters manuscript does not consist entirely of verses and of formulaic depictions of elephants, many of them unfinished. The most beautiful illustrations and the most intriguing content are to be found instead in the opening pages. Immediately preceding Chaddanta (Figure 1.1) are the three celestial elephants (Figures 1.3–1.5), which do form a part of the *Characteristics* text; before them are the three presiding deities, which do not. The first divine elephant (Figure 1.3) is composite—an Indian device, seen in a number of variations in Indian manuscripts. The accompanying stanza can be translated as:

An assembly of gods / and goddesses
Twenty-six in all come / taking their places
Forming a pachyderm / that's pure albino
Gods exclusively gather on high / appearing as an elephant

First in the group are the planets, which are also the days of the week, and the remaining gods, including the Earth and Nāga kings, come from different parts of the cosmos.[11]

The second of the three divine elephants is Erāvaṇa, pronounced Erawan in Thai (Figure 1.4). Erawan is the mount of the god Indra, who in Buddhist cosmology dwells in a palace on the summit of Mount Meru, the cosmic mountain. The third divine elephant is Girimekhalātraitayuga (Figure 1.5). The name is seldom encountered, but this elephant is the mount of the devil Māra, who attacked the Buddha at the time he attained enlightenment. He figures prominently in Thai mural painting, in the scenes of Māra's attack that appear on the entry wall of Thai image halls.

In the library of the abbot of Wat Phra Chettuphon, Prince Paramanuchit, a learned scholar, historian, and poet who died in 1851, the elephants can be viewed spatially (rather than in the sequential order of a manuscript).[12] Wat Phra Chettuphon, commonly known as Wat Pho, is adjacent to the royal palace in Bangkok, and the library dates from the reign of King Rama III (1824–1851). It is not the main temple library but a library for the abbot's private use, situated in the area set aside for the monks' dwelling places. It is raised on a high foundation and ornamented in a diaper pattern of colored glass. The entrance door is on the west; on the north and south walls there are three windows, and there are two on the east. On the east wall, which we face as we enter, above the two windows are the three celestial elephants. On the north, to our left, the mundane sequence of elephants begins, with Lake Chaddanta, in the *Himaphan*, depicted as a large round pond. The elephant Chaddanta himself stands beside it on a platform. Originally painted silver, he has tarnished. This is a case when the words of the poem, "as clear as beautiful silver," were taken literally. Elephants from the Brahma, Shiva, Vishnu, and Agni families are arrayed around the walls. The elephants of the ordinal and cardinal directions are

Figure 1.3. Composite elephant, first of the three celestial elephants. Walters Art Museum W.893, face 1, fol. 6.

Figure 1.4. Erāvaṇa (Erawan), mount of the god Indra. Walters Art Museum W.893, face 1, fol. 7.

Figure 1.5. Girimekhalātraitayuga, mount of the devil Māra. Walters Art Museum W.893, face 1, fol. 8.

placed in the appropriate positions. The inauspicious elephants, described in the last section of the poem, are clustered on the southwestern corner. On the south wall, the two main panels are devoted to depictions of the Chaddanta birth story.

As we face the east wall, we see the composite elephant at top left, Erawan at top center, and Girimekhalā on the right. Erawan, therefore, with his thirty-three heads and bearing the howdah for the god Indra on his back, is placed along the central axis—appropriate for an elephant and a deity associated with the cosmic mountain, Meru. Extending the presentation of the cosmological order is the presence along the top of the walls of small disks, which are symbols of the twenty-seven constellations of the ecliptic. Numbers 20 through 26 or 27 appear on the east wall.

The three presiding deities that appear in the manuscript (as will be seen) can be connected to ceremonies carried out by the court Brahmans of Siam, both in Ayutthaya (from 1351) and Bangkok (from 1782).[13] These deities do not appear in the library. It is recognized that their sphere is properly not that of Buddhist monks. But Brahmans and Buddhists share the elephants—closely, but not exactly, mainly because some texts assign different names and characteristics to certain elephants. We can say that the library is a Buddhist space because of the fact that the *Himaphan* was the setting not only of the Chaddanta birth story but of many other tales of the Buddha's previous lives as well. It was, therefore, a place for the accumulation of Buddhist virtues. What is equally important is the cosmological ordering; the library is an arena in which all the elements find a place in accordance with the Buddhist conception of the universe.

It is not known whether there were earlier buildings with murals organized according to the same principles. Since the list of pachyderms includes elephants assigned to the eight directions, it seems quite possible. This affects our understanding of the manuscript. It is linear, but anyone who is familiar with this library, when perusing the manuscript, will be able to envisage the elephants arrayed spatially.

Function

The elephant manuscripts can be said to have been emblems of office for officials involved in the management of elephants—the capture, domestication, and training of elephants for use in warfare, transportation, and industry. The information they contain is the esoteric counterpart, it might be said, of the practical knowledge required for the control of elephants. Still, this "esoteric counterpart" is not exactly a realm apart because the real-life skills involve a vast body of knowledge of magical texts and of procedures involving offerings. Capturing an elephant is inconceivable without the recitation of mantras, or magical sayings.

The concept of an "emblem of office" is in part based upon the information provided by the German traveler and ethnologist Adolf Bastian, who was in Bangkok in 1863 when a white elephant was discovered in the forest and brought to Bangkok. Bastian wrote,

> To receive him, the king traveled towards the elephant for several days, and in Bangkok a richly gilded stand was set up in front of the palace gates for the animal's arrival, on which the elephant, served

> by kneeling princes and nobles, was displayed for several days to the people, who also reveled in the puppet theater and fairground stalls that appeared. Beside the elephant, which was wearing a gilded harness and which rocked back and forth under a white baldachin, a seat covered with carpets was set up for the king, who was carried there on a sedan with a silver footrest. Gold and silver trees were erected as a sign of homage. The main role in this ceremony was played by a younger brother of the king, whose duty as marshal of the kingdom's elephants was to administer all matters relating to elephants. He kindly allowed me to borrow a book depicting and describing all the different species of elephant, so that one could derive an animal's pedigree from the characteristics listed and thus assess to what degree it was thoroughbred.[14]

In subsequent paragraphs, Bastian provided a careful description of the manuscript he borrowed, and it is clear that it is nearly identical to the manuscripts now at the Walters and elsewhere. In 1863, the elephant marshal was a prince, though some holders of this position must have belonged to nonroyal families and inherited the post.

Although Bastian understood the book to be a practical handbook, it is not really. Obviously, an elephant manager needs to have considerable practical knowledge. What the *Treatise on the Characteristics of Elephants* presents are tutelary deities and poetic literature that belong on a more elevated plane. The manuscript serves as a kind of badge of authority. Like a medal worn by an army officer, it is a recognition of the mastery necessary to oversee elephants. Elephants needed to be captured in the wild; they had to undergo rigorous training; they needed to be maintained; and they were essential in warfare and transportation.

The festivities described by Adolf Bastian were connected with the arrival of a white elephant in Bangkok. White, or albino, elephants are identified by specific traits that are not described in the poetic "Elephant Characteristics" texts. They were considered to be of divine ancestry, and their possession by the king was a sign of his accumulation of Buddhist virtues. A number of manuscripts date from the 1810s, during the reign of King

Rama II. By the end of his fifteen-year reign, the king had acquired six white elephants, each of whom was given a fancy name. This is the name of the first one: "*Phrayā* śvetakuñjara atiśara *prasœt* sakti *phüak* eka-argayairā maṅgala-bāhana-nārtha parama-rāja-cakrabarti *wichian* ratana-nāgendra jāti-gajendra Chaddanta hirañya-raśmī-śrī *phra* nagara sundhara-lakṣaṇa *lœt fā*."[15] The words that can be considered native Thai have been italicized, and loanwords from the classical language of India, Sanskrit, have been left in roman type. One name is worthy of notice: Chaddanta, as has already been indicated, was a bodhisattva, a Buddha-to-be, one of the incarnations of the Buddha before he was born on Earth in the fifth or fourth century BCE, and this name gives to the white elephant an especially Buddhist character. King Rama II in fact accumulated fewer white elephants than either Rama I or Rama III. Nevertheless, it is easy to imagine a sequence of ceremonies celebrating the arrival of a white elephant and the need for the officials responsible to be presented with a manuscript like the ones in the Walters and elsewhere.

Another interesting text dates from the reign of Rama II. In 1812, a white elephant was discovered near Battambang (today, in western Cambodia), and a written account survives of its discovery and of the path of its progress as it was led to Bangkok.[16] In 1927, this text was published in a printed edition, and it was remarked that it could serve as a model for the protocol to be followed when a white elephant is brought to the capital. This was Rama II's first white elephant, the one given an excessively lengthy proper name.

The Three Presiding Deities

Following introductory verses that mention the four elephant clans, the very first illustration appears, representing a break in the poem. The upper part of this opening (Figure 1.6) provides the words of a mantra, in Sanskrit; on the lower part is an explanatory text in Thai. We recognize here the Indian god Ganesha, the elephant-headed son of the great Hindu god Shiva. Ganesha, "lord of obstacles," is worshiped in India at the start of

Figure 1.6. Śivaputrabighneśvara (Lord of Obstacles, Son of Shiva), first of the three presiding deities. Walters Art Museum W.893, face 1, fol. 3.

an enterprise. Actually, this deity is a Thai manifestation of Ganesha who seldom appears outside the elephant manuscripts. He holds a trident in one hand (alluding to Shiva) and a lotus in the other.

Understanding the significance of this deity is possible by making use of three different kinds of evidence: first is a narrative that is found in an early Bangkok-period text called "Incarnations of the Gods"; the second consists of the words on this page of the manuscript; and the third is a recent analysis of the different branches of traditional elephant lore.[17] This Ganesha is described in the "Incarnations of the Gods," where he is given the name Śivaputrabighneśvara, Lord of Obstacles, Son of Shiva. Shiva gives a command to Fire (not called Agni but *phra phlœng*, "Holy Flame," the Angkorian *vraḥ vleng*) to produce two sons. In the flames that emerge from Fire's right ear appears Śivaputrabighneśvara, holding a *trīśula* (trident) in his right hand and a lotus in his left. From Fire's left ear comes

Koñcanāneśvara-śivaputra, who is the deity depicted on the following opening in the "Elephant Characteristics" manuscripts. The text on this page also tells us to think of the god and to repeat the mantra: that this elephant is the "platform" (*chœng*) for other elephants; that anyone who acquires this *gajaśāstra*, "elephant science," should pay homage to his instructors; and that prosperity in matters relating to elephants will result. (The word *chœng*, in Thai, is a loanword from Khmer.)

A book published in 2002 clarified Thai thinking concerning different texts and types of knowledge. *Gajaśāstra*, "elephant science," is the inclusive category.[18] It has two branches: *Gajalakṣaṇa*, "elephant characteristics," and *Gajakarma*, "elephant activities." *Gajalakṣaṇa*, "elephant characteristics," we have already learned something about; it is knowledge, represented on an elevated plane by the words of the "Characteristics" poem. *Gajakarma*, "elephant activities," will become clearer in the following pages, but the presence of the mantra here provides indications: physical activities involving elephant management are accompanied by the recitation of mantras. The analysis can be put in the form of a diagram (see below).

On the following opening (Figure 1.7) appears the divinity standing for "elephant characteristics." He is the god born from Holy Flame's left ear, and his long name is Koñcanāneśvara-śivaputra, Lord Elephant Head, Son of Shiva. If the most common Thai orthography is retained, this name, following Neelakantha Sarma, can be interpreted as derived from Sanskrit Kauñjarānaneśvara, "the lord with the elephant head," by way of Tamil Kauñcarānaneśvara.[19] There is no mantra on this page, and the text is explanatory. Koñcanāna's attributes are described precisely in the text "Incarnations of the Gods." In the upper two hands are two of the three divine elephants, who dwell in heaven rather than in the Himalayan forest. One is Erāvaṇa, the mount of the god Indra; the other is Girimekhalātraitayuga, the mount of the devil when he threatened the Buddha on the night of the enlightenment—both of which have been described already (Figures 1.3 and 1.5). In the middle hands are the male and female progenitors of the three grades of white elephant that will become the mounts of kings. White elephants, therefore, are of divine origin. In the lowermost hands are conches, evidently to be blown on ceremonial occasions.

Figure 1.7. Koñcanāneśvara-śivaputra (Lord Elephant Head, Son of Shiva). Walters Art Museum W.893, face 1, fol. 4.

The third deity (Figure 1.8) requires a much lengthier discussion. The caption at top center says, "This lord is called Debakarrma" (Devakarma)—pronounced *Thepphakam* in Thai. The suffix *karma* is common in texts having to do with elephants. In general, *karma* in this usage should be understood as meaning "business," "activities," or "affairs." *Gajakarma*, "elephant affairs," includes capturing and training as well as knowledge of mantras and of ritual activities. But the term also refers to personifications. This divinity, Devakarma, is a god of elephant affairs. On the far left, the caption reads "Ṛṣī Darabhāṣa Debakarrma," and on the right, "Ṛṣī siddhi braḥ karrma." Below, the two seated figures are labeled "Devaput," son of a god.

At lower left are five Sanskritic lines that probably could be called *wet* in Thai, that is, the Veda. The first two lines seem to be directed to Śivaputrabighneśvara (Ganesha), the first of the three presiding deities in the manuscript. He is addressed as *Mahodaro*, "great-bellied one," and his trunk achieves victory by means of loud trumpeting. The lasso, called a

Figure 1.8. Top center: Six-armed Nārāyaṇa Devakarma. Walters Art Museum W.893, face 1, fol. 5.

nāgapāśa, is invoked on the third line. The epithets in the fourth line appear to characterize a marauding elephant. The fifth line expresses hope for success.[20]

On the right-hand side of this opening are two lines with the following mantra:

> Om he he ṯiṣa caḥ ṯiṣa ha na ya
> Ṯejena asai svā hāya svā phat

And below this there are instructions in Thai with the meaning, "After the mantra is read three times the coiled and subjugated lasso becomes loose."[21] In other words, the power of the mantra and the power of the noose are identified with each other.

Fortunately, there is a text that explains at least some of this, as noted by Henry Ginsburg in *Thai Manuscript Painting*—the text referred to here as "Incarnations of the Gods" (*Thewa pāng*).[22] A group of wizards and hermits calls on Shiva on Mount Krailāsa to complain about an evil marauding elephant named Ekadanta, "one tusk." (In India, Ekadanta is a name for Ganesha, but this Ekadanta numbers among the elephants in the "Characteristics" texts.) Shiva sends two of the group to Nārāyaṇa (that is, Vishnu), then lying on the cosmic ocean. "Nārāyaṇa Debakarrma" (=Devakarma; pronounced Narai Thepphakam) comes to Earth in six-arm form, in each hand a weapon, one of them a lasso formed by the *nāga* king, Anantanāgarāja. Indian texts would call this lasso a *nāgapāśa*. (This term appears in the Sanskritic text appearing in Figure 1.8—but not in "Incarnations of the Gods"—and a *nāgapāśa* is depicted in the hand of a thirteenth-century Khmer bronze Ganesha.)[23] Nārāyaṇa encounters four farmers, who are naturally astonished at seeing a man with six arms. The god asks the location of Ekadanta; the farmers know, but it is a long trip, one that includes a sea voyage. Finally, they meet Ekadanta; Nārāyaṇa makes use of his weapons and tells the four men to memorize a mantra called *phrütthabāt*. Eventually Nārāyaṇa uses his serpent lasso to snare Ekadanta's right foot, and he is able to return to the cosmic ocean. The four farmers are now in possession of the *phrütthabāt* mantra, which they teach to their sons, who become masters of *gajakarma*, "elephant affairs," and the same privileged knowledge is passed down for the benefit of posterity.[24]

On the opening (Figure 1.8), Vishnu appears at center top with six arms, but five of the weapons have been eliminated. Only two serpent lassos remain. At left and right are the two Ṛṣī, "hermits," sent by Shiva to call on Vishnu. The two figures below, who are called the "sons of gods," do not match up with the text, but I understand them to represent two of the four farmers, now in possession of the *phrütthabāt* mantra. The farmers, it could be said, have become semidivine because of their knowledge of the mantra, which is symbolized by the snakes that encircle them (one long and blue, six shorter and red). The statement at the lower right, "After the mantra is read three times, the coiled and subjugated lasso becomes loose," indicates that the lasso can now be used to snare Ekadanta.

The word *phrütthabāt* calls for some discussion. It appears that the official modern spelling—which also has a slightly different pronunciation, *phrütthibāt*—is not the historically correct orthography, which is that of two Sanskrit words joined together.[25] The first is *bṛddha* (=*vṛddha*), "old," or "old one." The second is *pāśa*, "lasso." If this were a real Sanskrit compound, the meaning would be "old lasso." But the words have been joined together in a Thai way. The meaning is "old Brahman priest specializing in the lasso." Attached to the court in Bangkok and, before that, in Ayutthaya were Brahman priests who carried out royal ceremonies, following a ritual calendar. Among these Brahmans, *phrütthabāt* was a special rank. When the "Incarnations of the Gods" speaks of the *phrütthabāt* mantra, we should understand that as the mantra received from the god Nārāyaṇa by the figures encircled by serpents on the manuscript page and subsequently passed down to the court Brahmans who recited it in ceremonies.

Symbols and meanings are multiplying. We have serpents that are lassos that are mantras. We need to look at real lassos in order to better understand the relationship of mystical beliefs to actual practice. And we need to see what we know about the court Brahmans called *phrütthabāt*.

Real and Mystical Lassos

In 1930 and 1931, two remarkable articles on the actual hunting of elephants appeared, written by Francis H. Giles, an Englishman who served as an official in the Siamese government.[26] Giles made it clear how practical know-how is accompanied every step of the way by strict discipline and by the propitiation of gods or supernatural spirits; magical incantations are recited and statements are made concerning the offerings being presented. In regard to the lasso of the elephant hunters of northeastern Thailand, he wrote,

> The lasso is housed in a building standing high and separated from the homes of the villagers, no woman is allowed to trespass within the precincts of this house. A fact stands forth with clear definition which is of importance and must be set down here, it is that the lasso is not protected by any particular spirit or possessed by one. It

> is the spirit itself, and it is for this reason that such honor is paid to it. There is no ceremony of propitiation in order to obtain its favor as in the case of spirits which inhabit or protect certain things. Prayers of supplication are offered up begging that the lasso spirit will use its strength in furtherance of the interests of the hunt.[27]

Whether these villagers were aware of the mystic identity of the lasso, the *nāga*, and a mantra is not known. Giles gave no indication that they were. The practices he described seem to have roots that long predate the early Bangkok period. Therefore, the core belief historically, it can be argued, was in the sacredness of the leather lasso. The connections with the *nāga* and a mantra enhance that intrinsic sacredness; they did not give rise to it.

Giles translated many of the invocations recited by the head of the elephant hunters, but no transcribed text has ever been published. There are more elevated texts that correspond to those of the hunters, however. One of these, printed in 1914, was written in the early Ayutthaya period by a poet from Sukhothai. Another manuscript in the National Library, Bangkok, has been described only recently. It consists of invocations in a language that the editor considered comparable to that of the Angkorian inscriptions, together with statements about specific offerings.[28] These texts can be called "Propitiation Texts."

The term *phrütthabāt* appears just twice in the "Incarnations of the Gods," once called a mantra, once a formula (*tamrap*).[29] If we look for this compound in the Royal Institute dictionary (where it is spelled *phrütthibāt*), we find the following definition: "a group of Brahmans who carry out ceremonies involving elephants."[30] In the Ayutthaya chronicle compiled in the seventeenth century, an official is said to have served as the *phrütthabāt* in an elephant ceremony during the reign of King Chakkraphat (1548–1569).[31]

Phrütthabāt brahmans also appear in a passage in *The Chronicles of Ayutthaya* compiled in the early Bangkok period:

> During the year of the cock, ninth of the decade [1657 CE], the king [I will omit his long title] issued a holy proclamation command-

> ing Phraya Chakri to prepare a building for the Banchi Brahma ceremony and an assembly hall for the performance of all the royal ceremonies in the vicinity of the Elephant Corrals. And the king ordered [an image of] a golden Thewakam (= Skt. *Devakarma*), standing, one *sòk* high and gilded, the attire inset with gems [?], to be reserved for the ceremony of Khotchakam (= *Gajakarma*, "Elephant rites"). He had the *braḥ rāja grū phrütthabāt* (the royal preceptors specializing in the lasso) and the deputy royal preceptors perform the ceremony of Banchi Brahma on Thursday, the tenth day of the waxing moon in the seventh month. Thereupon, His Majesty, having proceeded to the rites of the Ceremony of the Sowing of the Heart and the Banchi Brahma Ceremony in the field of grass, the rites of the holy royal ceremonies were performed in every detail following the texts (*śāstra*) which exist for *Gajakarma*.[32]

The image in this ceremony is said to have been a "Devakarma," the same term used in the Walters manuscript to identify the six-armed seated Nārāyaṇa (Figure 1.8). He was more likely, however, a form of Ganesha that does not appear in the "Elephant Characteristics" manuscripts. There are numerous nineteenth-century iconographic compendia, evidently modeled on ones produced in southern India, and in these, "Devakarma" is consistently represented as a standing elephant-headed deity, holding a tusk in his right hand and a pole in his left.[33] That this icon had Ayutthaya-period predecessors is demonstrated by the existence of a bronze two-armed Ganesha in the Metropolitan Museum of Art, New York, dating from the late fifteenth or sixteenth century.[34]

The position *phrütthabāt* survived well into the nineteenth century. In his *Ceremonies of the Twelve Months,* King Chulalongkorn (Rama V, r. 1868–1910) described the role played by *phrütthabāt* Brahmans in the lasso ceremony of the fifth month, a ritual reenactment of the pacification of the elephant Ekadanta.[35] In the ceremony, homage is paid to one silver-plated and three gilded lassos. Indeed, according to the account of the capture and transport of the white elephant in 1812, "The lasso that had been used to snare the Lord White Elephant was lacquered and gilded and preserved as the premier lasso, for use in royal ceremonies."[36] This brings us full circle: *nāga* serpent = lasso = mantra = lasso used to snare a white elephant. On the village level, it has been seen, the lasso was a sacred object; here now, at the court level, it has become more numinous because the sacrality of the white elephant has enhanced it.

Origins, Structures

It is not yet possible to provide a full historical context for the mythology of the three presiding deities in the manuscript. Some aspects would appear to be inventions of the early Bangkok period. Only additional research—and speculation—will reveal the relative importance of earlier Ayutthaya practices, of connections with Cambodian traditions (leading back to Angkor), of the culture of the elephant hunters, and of the beliefs of Tamil Brahmans.

The very close connection between the myths recounted in "Incarnations of the Gods" and the depictions in the manuscripts suggest that at one point in the early Bangkok period court Brahmans systematized their beliefs. The fact that the introductory deity, Śivaputrabighneśvara, is generally absent in the miscellaneous iconographic compendia provides evidence for thinking that he was an invention rather than a well-established manifestation.[37] That Brahmans who had southern Indian origins had a significant role in this process is suggested by the Tamil element in the name Koñcanāneśvara-śivaputra.[38] As for the third deity, the six-armed Nārāyaṇa Devakarma, it remains to be determined whether he had Vishnu predecessors in the Ayutthaya period or whether he was a Bangkok-period replacement for an earlier Ganesha.

The triadic structure of these three deities points to various connections, but the systems are not entirely congruent. First, as indicated above, there is the structure of the content:

Śivaputrabighneśvara ("Lord of Obstacles, Son of Shiva")
Gajaśāstra, "elephant science"
Chœng (platform, base)

Gajalakṣaṇa "elephant characteristics"	*Gajakarma* "elephant activities"

Second, there is the myth presented in "Incarnations of the Gods":

Holy Flame, the Sacred Fire (*phra phlœng*),
as ordered by Shiva

From Fire's left ear Koñcanāneśvara-śivaputra	From Fire's right ear Śivaputrabighneśvara[39]

Third, in his account of elephant hunting, Francis Giles described three sacred fires that burned in front and to either side of the jungle hut of the leader of the expeditions. At these fires, the leader invoked the blessing of the god Agni and recited a prayer for success in taking elephants.[40] (There are also references to fires of the left and the right and to the god Agni in the early Propitiation Texts.)[41] The fire in front was the *chœng:*

Agni
Chœng (platform, base)

Fire of the left	Fire of the right

These three triadic structures bear a relationship, one to the other. But tracing each back and formulating a plausible historical reconstruction of the developments are tasks that lie ahead and that may be no easier than once was the capture of the elephants in the forests.

Notes

This chapter includes material presented in somewhat different form at a symposium at the University of Pennsylvania, 22 October 2011; at a panel of the European Association of Southeast Asian Archaeologists, Fourteenth International Conference, Dublin, 18 September 2012; and in a lecture at the Chester Beatty Library, Dublin, 20 September 2012. For assistance of various kinds, I thank Samerchai Poolsuwan, Phillip Scott Ellis Green, Justin McDaniel, Nandana Chutiwongs, Jana Igunma, Tom Patton, Arlo Giffiths, David Chandler, and the staff of the Chester Beatty Library and the Asia Reading Room of the Library of Congress. Thai words are represented in two distinct ways: phonetically, following a modified Library of Congress system; and orthographically (generally, only in the case of Sanskrit or Pali loanwords), using the standard Indic system.

1 Henry Ginsburg, *Thai Manuscript Painting* (Honolulu: University of Hawaii Press, 1989), 33–43; Ginsburg, *Thai Art and Culture: Historic Manuscripts from Western Collections* (London: British Library, 2000), 130–31; Ginsburg, "Thai Painting in the Walters Art Museum," *Journal of the Walters Art Museum* 64/65 (issue year 2006/2007, published 2009), 99–148 (136–37); *Samut Khòi* (Bangkok: Khrong kān süpsān mòradok watthanatham; distr. Khrusaphā, 1999), 78–115; Natthaphat Čhanthawit, "Chāng mongkhon," in *Sārānukrom watthanatham Thai phāk klāng*, 15 vols. (Bangkok: Mūnithi Sārānukrom watthanatham Thai, 1999), 4:1898–1919.

2 Natthaphat, "Chāng mongkhon" (note 1), 1912–1919.

3 "Tamrā laksana chāng kham khlōng," in *Tamrā chāng phāk thī 1* (Bangkok: Krom Sinlapākòn, 1938) (University of Michigan microfilm, MiU LP 0076.21).

4 *Kham chan Khotchakam Prayūn* appears in *Kham chan Dutsadī Sangwœi Kham chan Klòm Chāng khrang krung kao læ Kham chan Khotchakam Prayūn* (Bangkok: Fine Arts Department, 2002), 77–96, with a discussion by the editor, Bunthüan Sīwòraphot, 49–52.

5 In the printed text of "Kham chan" (note 4), p. 3. This and other translations made by the author no doubt include errors of understanding.

6 The Chaddanta Jātaka is number 514. The translated version is E. B. Cowell, ed., *The Jātaka; or, Stories of the Buddha's Former Births*, 6 vols. (repr. London: Pali Text Society, 1973), vol. 5, trans. H. T. Francis, 20–31.

7 For an illustration, see Ginsburg, *Thai Art and Culture* (note 1), 130.

8 Franklin Edgerton, *The Elephant-Lore of the Hindus: The Elephant-Sport (Mātaṅga -līlā) of Nīlakaṇha* (New Haven: Yale University Press, 1931), 55.

9 Ginsburg, *Thai Manuscript Painting* (note 1), 33.

10 S. Gopalan, ed., *Aśvaśāstram by Nakula with Coloured Illustrations* (Tanjore: T. M. S. S. M. Library, 1952), 215.

11 As described in Bunthüan, ed., *Kham chan* (note 4), 123.

12 *Hò trai krom somdet phra Paramānuchit Chinōrot: Læng rīan rū phra phutthasātsanā læ phumipanyā thai nai Wat Phō* (Bangkok: Khanasong Wat Phra Chettuphon, 2005).

13 H. G. Quaritch Wales, *Siamese State Ceremonies: Their History and Function* (London: Bernard Quaritch, 1931).

14 Adolf Bastian, *A Journey in Siam (1863)*, trans. Walter E. J. Tips (Bangkok: White Lotus Press, 2005), 72–73.

15 The article "Chāng," in *Sārānukrom watthanatham Thai phāk klāng* (note 1), 4:1885.

16 *Čhotmāi het rüang rap phrayā sawettakunchòn chāng phüak ræk dai nai ratchakān thī 2* (Bangkok: Watchīrayān Library, 1927).

17 "Incarnations of the Gods": *Thewa Pāng* (Bangkok: Krom Sinlapakòn, 1970). I thank Tom Patton for making a PDF of the copy in the Cornell University Library. The same content can be found in the following three publications: "Tamrā Khotchakam," excerpted from a manuscript of the same name (catalogued as no. 108 in the National Library), in Bunthüan, *Kham chan* (note 4), 171–74; the introductory section of the text *Nārāi sip pāng*, first printed in 1874, from a printing under the title *Nārāi sip pāng læ phong nai rüang Rāmakīan*, which appeared in 1967 (this was the edition consulted by Henry Ginsburg; I thank Jana Igunma); and a text by King Mongkut, excerpted at length in *Samut Khòi* (note 1), 81–97, and printed also in *Chumnum phra bòrommarātchathibai nai phra bāt somdet phra Čhòm Klao čhao yu hua, muat wannakhadī læ muat bōrānnakhadī læ prachum phra rātchaniphon nai ratchakān thī 4 phāk pakinka* (Bangkok, 1929).

18 Bunthüan, ed., *Kham chan* (note 4), 25.

19 Neelakanta Sarma Pandit, "Un album thaïlandais d'iconographie indienne," *Arts Asiatiques* 26, no. 1 (1973): 157–89 (160).

20 The following translation of this text is almost entirely the responsibility of Phillip Scott Ellis Green, to whom I express my gratitude:

(1) *Mahodaro*, Big-bellied one! *Maḥhākāyo* (= *mahākāyo*), Great-bodied one! *Śivapuṯṯro*, Son of Śiva!

(2) *Mahidriko*, Powerful one! *Hatthācāro* (read *hatthācaro*?), One whose trunk is immoveable! *Hatthijayochcā* (read °*uchcā* as °*uccā*?), Possessing a trunk (that achieves) victory by means of loud noise!
(3) *Hi caḥ* (no meaning). *Pāśadharo*, Holder of the noose! *Nāgapāśo*, The *nāga* noose! *Nāgabăndho*, The *nāga* snare!
(4) *Gajarăksoca* (read *gajarakṣasa*), Elephant demon! *Acai* (*ācai* is perhaps related to Tamil *ācai*, desire, lust, passion). *Jabbaṯeje caḥ* (= *jambha tejas*), Sharp-jawed one!
(5) *Dverājā*, Twice king! *Trīyadevā*, Three gods! (?). *Caḥ me siddhi*, Success to me! (If the translation is correct, this is the most grammatically incorrect line. Maybe *dverājā* is an outgrowth of *dvija*, twice-born, and refers to Gaṇeśa. "Three gods" might be the three presiding deities of the manuscript. For the general sense compare, in Bali, the final line of a hymn directed to Gaṇeśa, *siddhir bhavatu me sadā*, or "Let success always happen to me," T. Goudriaan and C. Hooykaas, *Stuti and Stava (Bauddha, Śaiva and Vaiṣṇava) of Balinese Brahman Priests* [Amsterdam: North-Holland, 1971], 26. Also *Siddhir bhavatu me Deva tvat prasādāt Maheśvara*, or "O Deva, O Maheśvara, by Thy grace, may success be mine," as found in Srisa Chandra Vasu, *The Daily Practice of the Hindus, Containing the Morning and Midday Duties* [New York: AMS, 1974], 150.)

21 This brief text, มนตนี้ อ่าน ๓ คาบ คัด บาศฏูกย่อมลุ่ยแล, is not easy to translate. All three *mai ek* tonal markers appear to be emendations. The first part is straightforward: "Read this manta three times." Subsequent words vary somewhat, manuscript to manuscript: the Chester Beatty Library manuscript (where all the words appear on a single line) has ยอม and ลุย rather than ย่อม and ลุ่ย. There may be other variations in a manuscript in Bangkok (http://watthakhanun.com/webboard/showthread.php?p=14939&langid=1, accessed 9 November 2013). For clarifying the meaning I am grateful to Dr. Samerchai Poolsuwan of Thammasat University. Following his analysis, I take *khat bāt* as an inversion of *bāt khat* (as in Sanskrit compounds), meaning that the subject is *pāśa*, "lasso." The following word, *thūk*, goes with the previous two and implies an unstated *mon* (*mantra*): thūk mon, "to be under the spell of a mantra." Hence, "the coiled and subjugated lasso."

22 Ginsburg, *Thai Manuscript Painting* (note 1), 36–38. See note 17 above.

23 Emma C. Bunker and Douglas Latchford, *Khmer Bronzes: New Interpretations of the Past* (Chicago: Art Media Resources, 2011), fig. 9.54, p. 436; text, p. 425 (National Museum of Cambodia Ga 2320).

24 *Thewa Pāng* (note 17), 8–11.

25 The official spelling is *bṛdhipāśa* (see *Photčhanānukrom chabap Rātchabanditsathān phò sò 2492* [Bangkok, 1964], 645), defined as "a group of Brahmans who carry out ceremonies involving elephants." *Bṛdhipāśa* appears not to be a valid Sanskrit compound. In printed texts, manuscript spellings have been altered to conform to this spelling. Older manuscript spellings include *bṛdipāda* (pronunciation *phrütdibāt*) and *bṛddhapāśa* (pronunciation *phrütthabāt*), both in the *Chronicle of the Kingdom of Ayutthaya: The British Museum Version* (Tokyo: Centre for East Asian Cultural Studies for UNESCO, The Toyo Bunko, 1999), 39 (fol. 37v) and 323 (fol. 319r). In deciding upon the spelling *bṛddhapāśa*, I am grateful for the opinion of Nandana Chutiwongs. For *bṛddha* as "old" or "old man" in Thai and Khmer, see George Bradley McFarland, *Thai-English Dictionary* (Stanford, CA: Stanford University Press, 1972), 570; Saveros Pou, *Dictionnaire vieux khmer-français-anglais* (Paris: L'Harmattan, 1992), 467; Joseph Guesdon, *Dictionnaire Cambodgien-Français,* 2 vols. (Paris: Plon, 1930), 2:1263–65. Khmer inscription K. 569 (Banteay Srei, 1304 CE) includes the proper name, "K. A. Paṇḍit Vṛddhācārya." See Louis Finot, H. Parmentier, and Victor Goloubew, *Le temple d'Içvarapura (Bantay Srei, Cambodge)* (Paris: G. Vanoest, 1926), 77, 81. An alternative possibility (for which support in Thai sources has not yet been found) is *vṛttapāśa*. Dr. Arlo Griffiths has written me (e-mail, 26 July 2012) as follows: "The compound *vṛttapāśa* would mean 'wielding/applying the lasso,' and this plausibly yields a derived meaning, 'carrying out ceremonies involving elephants.'"

26 Francis H. Giles, "Adversaria of Elephant Hunting (Together with an Account of All the Rites, Observances and Acts of Worship to Be Performed in Connection Therewith, as Well as Notes on Vocabularies of Spirit Languages, Fake or Taboo Language and Elephant Command Words," *Journal of the Siam Society* 23, pt. 2 (1929–30): 61–96; Giles, "An Account of the Rites and Ceremonies Observed at Elephant Driving Operations at the Seaboard Province of Lang Suan, Southern Siam, *Journal of the Siam Society* 25, pt. 2 (1932): 153–214.

27 Giles, "Adversaria" (note 26), 76.

28 *Khamchan Dutsadī Sangwœi*, first printed in 1914, is discussed in Bunthüan, ed., *Kham chan* (note 4), where the text appears, pp. 55–64. Excerpts from the new text, taken from a National Library manuscript numbered 99 and titled *Tamrā Phra Khotchakam*, appear in this book as well, and the editor, Bunthüan Sīwòraphot, discusses the dating, pp. 25–30.

29 *Thewa Pāng* (note 17), 10.

30 See note 23 above.

31 Richard D. Cushman, trans., *The Royal Chronicles of Ayutthaya* (Bangkok: Siam Society, 2000), 28; "Phra Kamwaca [*Kammavācā*] as the . . . Phrüetthibat." See also note 23.

32 Translation adapted from Cushman, trans., *Royal Chronicles of Ayutthaya* (note 31), 245. His "of Royalty niello": maybe *thom* means "filled" (although there may have been *niello* in Siam, as it existed in India at this time). Maybe *wæn* is not "ring" but ring-shaped settings for gems. For the *bṛdhipāśa* of the printed text (see note 23 above), he has "Snares of Plenty." I do not know the meaning of *banchī* in "Banchi Brahma"; Cushman has "the Register of Brahma." For the Thai text, see note 23 above, and *Phra Rātchaphongsāwadān krung Sayām čhāk ton chabap khòng Baritit Miusīam* (Bangkok: Kao Na, 1964), 387.

33 *Tamrā phāp thewarūp læ thewadā nopphakhrò* (Bangkok: Krom Sinlapākòn, 1992), 28, 53, 191, 192. A manuscript nearly identical to the last of the five in this publication, in the École française d'Extrême-Orient, Paris, was published with valuable identifications in Sarma, "Un album thaïlandais d'iconographie indienne" (note 19). The tusk identification is made by Bunthüan Sīwòraphot in *Kham chan* (note 4), 35.

34 Metropolitan Museum of Art 1983.512. Forrest McGill, ed., *The Kingdom of Siam: The Art of Central Thailand, 1350–1800* (San Francisco: Asian Art Museum, 2005), 139–40.

35 Chulalongkorn, King of Siam, *Phra rātchaphitthī sipsòng düan* (Bangkok: Khlang Witthayā, 1964), 249–54. A brief description of this ceremony will appear in a forthcoming article by the author.

36 *Čhotmāi het rüang rap phrayā sawettakunchòn chāng phüak ræk* (note 16), 36.

37 For an exception, see *Tamrā phāp* (note 33), 64.

38 For the Indian origins of the Siamese Brahmans, see Jean Filliozat,"Kailāsaparaṃparā," in *Felicitation Volumes of Southeast-Asian Studies Presented to His Highness Prince Dhaninivat Kromamun Bidyalabh Bridhyakorn*, 2 vols. (Bangkok: Siam Society, 1965), 2:241–48.

39 Because Śivaputrabighneśvara is called Khandakumāra (interpretable as the Hindu god Skanda, Shiva's other son) before he receives the elephant head (*Thewa Pāng* [note 17], 7), one line of approach would be to see one of the two sons of Shiva as like the Indian Skanda, the other like the Indian Ganesha. For an approach along these lines, G. E. Gerini, *Chūḷākantamaṅgala: The Tonsure Ceremony as Performed in Siam* (Bangkok: Siam Society, 1976; orig. publ. 1893), 15–17. There are difficulties, however, with such an interpretation.

40 Giles, "Adversaria" (note 26), 85. "The fire at the right and left side of the hut are called 'Khampuak-sadam' (right) and 'Khampuak-sadian' (left), that at the front, 'Khampuak-churng' (foot)." *Sadam* is the standard Khmer word for *right. Khampuak* has resisted explanation.

41 Bunthüan, ed., *Kham chan* (note 4), 38–39.

CHAPTER 2

Representations of Space and Place in a Burmese Cosmology Manuscript at the British Museum

Alexandra Green

Like textual descriptions of the cosmos,[1] Burmese cosmology manuscripts are pragmatic, focusing on the universe's spatial arrangements visually and providing information about the characteristics of various locations in the cosmos. They are also highly standardized, in keeping with Burmese artistic production in the eighteenth and nineteenth centuries. Usually, the tripartite cosmos, comprising the realms of formlessness, form, and desire, is structured vertically and centered on Mount Meru on one side of the manuscript. Encircling Meru are seven mountain ranges, around which are celestial bodies, including the sun and the moon represented by a peacock and a rabbit, respectively.[2] The second side comprises the Manussa, or human, realm and scenes of the Hells. The levels of the universe within the three depicted regions (Heavens, Hells, and the human realm) are described textually with factual details, such as the distances between areas and the qualities of the beings who are born in each realm. Visually, the actual structure of the cosmos is maintained as much as possible, although the section on the Manussa realm includes additional material that focuses attention on enlightened beings and Himavanta.

Cosmology in the form of Mount Meru and the surrounding mountain ranges is a regular, although not highly frequent, theme in Burmese art. One of the earliest representations occurs in the wall paintings of the Lokahteikpan temple at Pagan that date to the twelfth century. These murals depict Mount Meru as a pillar flanked by a cross section of the seven mountain ranges to each side, also represented as pillars, with giant fish swimming in the oceans at the base. Each pillar is topped with a deity in a building, and the pillars themselves are decorated with a red, white, and black pattern of stylized rocks that may be the source of the honeycomb pattern ubiquitous on later representations of the pillar-mountains. On the right side, the Buddha ascends toward Tāvatiṁsa Heaven located at the summit of the Meru pillar, and on the left, he descends back to Earth accompanied by numerous gods, including Indra (Sakka) and Brahma. In the skies around the mountains are the sun, moon, and various planets. In Tāvatiṁsa Heaven, the Buddha sits in a large, ornate wooden palace surrounded by devotees, lotus ponds, and two stupas, indicating the worship of his relics. The portrayal at the Lokahteikpan was emulated in later artistic production, including as a large scale, sculpted model at the Powindaung Caves, where it is over ten meters high, and in Nyaungyan and Konbaung dynasty (seventeenth through nineteenth century) murals.[3] In 1816, a three-dimensional model of the universe, the Hsinbyume stupa, was produced in Mingun as a memorial upon the death of the future King Bagyidaw's chief queen. In this example, the central stupa was set high upon an artificial hill and encircled by receding levels, representing the mountain ranges and oceans surrounding Meru around which devotees could circumambulate. There are also other cosmological representations. For instance, a few mural sites from the Pagan period and later display circular images of Lake Anotatta and the four major rivers that spring from it. Although not as extensively produced as other manuscript forms, cosmology manuscripts that date from the eighteenth and nineteenth centuries are extant in a number of libraries and museums.

Representations of cosmology are a materialization of ideas about the functioning of the universe, particularly the law of cause and effect,[4] and it is the relationship between karma and a being's position in the universe that is significant in representing cosmic structure. As Juliane Schober writes, "[People] . . . venerate and make offerings . . . to participate ritually in the continual manifestation of the Buddha at centers that hold his remains, or symbolic substitutes thereof and thus replicate a universal order in their local worlds and integrate their lives into an encompassing Buddhist cosmos."[5] Buddhist practice in Burma is characterized by an expectation of advantages (Pali: *ānisaṁsa*) resulting from appropriate behavior, with the desired results often expressed in wishes and prayers (Pali: *panidhana*). Merit generation contributes to advancement toward the ultimate goal of nirvana, with consequent, intermediary material benefits. The latter make merit levels visible to surrounding communities and therefore contribute to social structuring.[6] A cosmology manuscript not only encoded the structure and function of the universe, but it also provided the donors with social and religious status. In this way, cosmology manuscripts both reflect religious and social beliefs and reinforce those concepts visually.[7]

The British Museum Manuscript

In 2010, the British Museum purchased a Burmese cosmology manuscript, a *parabaik*, made of locally produced white mulberry paper with thick, green outer covers also of paper. It was acquired in Burma by Sir Frederick William Richard Fryer KCSI (Knight Commander of the Order of the Star of India), chief commissioner and then lieutenant-governor of Burma from the mid-1890s until 1903. Burma was his last posting in the colonial service, and he returned to Britain upon retiring. He must have acquired the manuscript during his tenure in Burma, yet given its condition—worn cover, used pages, repairs to the folds—it must have been made well before the purchase, although some of the damage may have been sustained during its years in the United Kingdom. The style of the painting is that of the second quarter of the nineteenth century, and the manuscript was probably produced around that period.

The format of the British Museum cosmology manuscript, combining illustrations and text, is in keeping with other known examples.[8] The writing is either interspersed among the pictures or arranged in text-only sections across several folios on both sides of the manuscript. There are large, blank spaces on a number of pages, and several sections have been covered with white-out and rewritten. While irregular arrangements of text and imagery are common, unusually some of the texts in the British Museum example are fragmentary. It is beyond the scope of this paper to delve into the possible reasons for the incomplete nature of the manuscript and the irregular organization, but the fact that the Buddhist universe was not the most popular or regularly produced subject matter may have affected production knowledge. It is also possible that apprentices were set to this work. This requires further research as there will undoubtedly be many other explanations for the odd disposition of material across cosmology manuscript pages.

The British Museum manuscript is a *nissaya* with alternating Burmese and Pali text written in Burmese script. The content of the text also accords with other, similar manuscripts in providing lists of information about, for instance, the eight dangers, the nine good qualities of a royal advisor or a minister at court, and children's duties to their parents, in addition to a description of the thirty-one abodes and the structure of the universe. The composition of Mount Meru and the names of important sites and their characteristics are detailed, as are facts about such features as the type, age, and height of the beings eligible to be born on each cosmic level and the trees that grow there. As with the text accompanying the Heavens, the writing that accompanies the Hells enumerates briefly the causes of rebirth in these realms. There are lists of the distances between the elements of the universe, and the manner in which each level will eventually be destroyed at the end of the *kappa* cycle is also detailed. The text does not always correspond with the imagery; at times it identifies what is depicted in the paintings, but at others it addresses different religious issues. Sometimes Pali texts are cited as source material for the information provided.

Visually, the British Museum manuscript relates to other late nineteenth-century examples in that on one side of the volume are schematic representations of the Heavens reaching from the pinnacle of the world of formlessness, the Heaven of Nevasaññā-nāsaññāyatana, down to the cosmic waters around Mount Meru and the realm of the Asuras (levels 1 to 26 and 28), and on the reverse side are images from the Manussa realm and scenes of torments in the eight levels of Hell (levels 27 and 31, respectively). Typically, levels 29 and 30, the worlds of the animals and ghosts (*peta*), are not illustrated.[9]

Imagery of Cosmic Structure

The representations comprise three distinct sequences, mostly arranged vertically. The first occupies the side of the manuscript with the Heavens and the realm of the Asuras and presents a coherent view of the cosmic structure. The scenes of Hell continue the list-like structure of the Heavens, while the images of the world of humans (Manussa) includes highly schematized diagrams of the four continents that surround Mount Meru and the sixteen *Mahājanapadās* (sites associated with the Buddha). However, there are also scenes of the Himavanta region, arranged both vertically and horizontally, illustrating the loveliness of the area and connecting it with enlightened beings. Each of these sections is portrayed in a manner that emphasizes specific concepts, notably the enumeration of cosmic structure and karma or an emphasis upon Gotama Buddha and his enlightenment.

Pictorially, the four levels of nonform are depicted singly as empty, undifferentiated wooden palaces with elegant carvings and embellishments typical of palace and monastic structures in Burma from at least the Pagan period (c. eleventh through thirteenth century). This addresses the issue of formlessness by representing the luxury and comfort and therefore the bliss of those beings who have attained these exalted levels of existence, yet without showing them bodily (Figure 2.1).[10] The palaces of the sixteen Heavens below these are populated, usually with a single figure, a *brahmā* deity, as befits the realm of form. The palaces of levels 5 to 9 are also shown singly, those of levels 10 and 11 are shown as a pair, and the palaces of levels 12 to 20 are grouped in threes. The palaces of Paranimmita-vasavattī and Nimmānarati Heavens (21 to 22) are shown singly again, as are those of Tusita, Yāmā, Tāvatiṁsa, and Cātummahārājika Heavens (23 to 26). This is a standardized format typical in representations of cosmological structure.

While there is usually little to differentiate the levels beyond the caption with the name of the Heaven below the palace pavilion, a few have been selected for more detailed portrayals in the British Museum example. On level 5, Akaniṭṭha Heaven, the Dussacetiya stupa built by Ghatīkāra Brahmā to house the discarded clothing of Prince Siddhattha, a *tagunda-*

Figure 2.1. The Ākāsānañcāyatana (level 4) and Akaniṭṭha (level 5) heavens. Burmese cosmology manuscript, late nineteenth century. 2010,3003.1. © The Trustees of the British Museum.

ing flagstaff, and an umbrella form are represented beside the palace (see Figure 2.1). Tusita Heaven, the realm where all future Buddhas are born in their penultimate life before becoming enlightened, has both a palace and the Sudhammā Zayat preaching hall. Tāvatiṁsa Heaven is always shown with a number of features because it is the home of Sakka (Indra; Burmese: Thagya-min) who is intimately connected with Burmese kingship,[11] as well as being the place where the Buddha preached to his mother. In the British Museum manuscript, this heaven is also the only one to include background decoration—deep blue at the top and bottom of the image and a green area in the center (Figure 2.2). The depiction of the heaven includes Sakka's Wejayanta palace, where he lives in luxury with his four queens; a *tagundaing* pillar and an umbrella next to the Cūḷāmaṇicetiya (Burmese:

Sulamani), the stupa that holds Gotama Buddha's hair relic; and another Sudhammā Zayat where the gods assemble and preaching occurs. There are also two gateways shown, and since in other manuscripts they are usually the Cittakūta and Suvaññacitta entrances,[12] the same can be supposed here. Less usually, there is also a depiction of the Buddha seated under a blooming tree, the Pāricchattaka, which grows at the foot of Sakka's stone seat. It is from this seat that the Buddha preaches to his mother, and in this instance, two identical *devas* (minor deities) pay homage to him, although whether one is his mother remains unclear. Most cosmology manuscripts also include the *padesa* wishing tree, which supplies whatever is desired, and some portray the sea coral tree, one of the five celestial trees. Neither is shown in the British Museum example. Below Tāvatiṁsa are Cātummahārājika Heaven and the start of the Meru pillar that is embellished with the traditional honeycomb pattern in green and white. Here, Cātummahārājika is represented by a single, beautifully decorated wooden palace that is much larger than the palaces of the other heavens and contains one of the kings of the four directions plus two women, probably his wives (see Figure 2.2). All are sumptuously dressed in clothing associated with the Burmese court, indicating their high status and luxurious living conditions. The additional stupa and preaching hall imagery in the Akaniṭṭha, Tāvatiṁsa, and Tusita Heavens is associated primarily with Gotama Buddha, either events in his life or his relics, yet the palace scenes of Tāvatiṁsa and Cātummahārājika Heavens illustrate the importance of kingship in Burmese Buddhist society by showing heavenly kings and thereby also demonstrating the benefits of good karma. The embellishment of Tusita Heaven also hints at the coming of future Buddhas, currently Maitreya.[13]

The next two openings display the mountain ranges in typical fashion as pillars with palaces at the summit (Figure 2.3). The sun, moon, and celestial orbs hang in the sky. On the slopes of Mount Meru are the realms of

---

Figure 2.2. The Tāvatiṁsa (level 25) and Cātummahārājika (level 26) heavens and the beginning of the cosmic pillar. © The Trustees of the British Museum.

the ogres, demons, *garuda* birds, and *nāgas* (mythical serpents); pictorially, these are arranged one above the other in elaborate settings on the central pillar. The British Museum manuscript includes a fifth group, that of the *gandabas*, celestial musicians, a feature that can be seen in a few other manuscripts. At the base of the pillar, the king of the fish, Ānanda, swallows his tail in an ocean full of smaller fish.

Separated from the ocean scene by a blank space, the next image depicts the realm of the Asuras, level 28 (Figure 2.4), instead of level 27, the human world, which is located on the reverse of the manuscript. This may result from a practical reason, namely, that the Manussa realm is composed of a number of scenes and images that spread over several folds and thus will not fit on the same side as the Heavens and the cosmic pillar, if the imagery were to remain in the current proportions. It may also result from the idea that the Asuras live within a hair's breadth of Mount Meru and that the Asuras once lived in Tāvatiṁsa Heaven but were cast out.[14] The less blissful existence of beings in the Asura realm is indicated by the plainness of the buildings and the clothing compared with the sumptuousness of the Heavens, as well as the presence of an image of distress—a seminude woman being bitten by a dog.[15] This is the only place in cosmology manuscripts where punishment is shown outside of the Hell regions, an indication of its spiritual deterioration.[16] Organizationally, the arrangement of this side of the cosmology manuscript is conceptually tidy, moving in a summary format from the exalted states of formlessness to the realm of form and then the base of Meru and the beginning of the states of woe. It thereby details in a single, extended visual sequence the realms within which beings can be reborn, and by implication tells of the law of karmic cause and effect by which the cosmos operates.

The illustrations on the second side of the cosmology manuscript include seven openings relating to the human world (level 27) and eight

---

***Figure* 2.3.** The cosmic pillar and waters. © The Trustees of the British Museum.

Figure 2.4. The realm of the Asuras (level 28). © The Trustees of the British Museum.

Figure 2.5. Roruva Hell, the screaming Hell where beings go if they have been miserly, committed adultery, or denigrated the *Dhamma*. © The Trustees of the British Museum.

depicting the eight regions of Hell (level 31), the latter in a repetitive format. Each of the Hells, displayed over a single fold, shows a cauldron packed with people being boiled, a common representation that is also found in Burmese wall paintings. There are also images of specific tortures related to particular crimes (Figure 2.5), with those who behaved worst placed in the lowest Hell. On each of the eight levels, the guardians of the Hells are seated in wooden buildings where deputies present them with information about people's evil deeds. As with the illustrations of the Heavens, the repetitive representations of the Hells are arranged in order as a list, with the details in the imagery used to reinforce ideas about the vertical format of the cosmos that ethicizes the universe with the ascent of those with good karma and the descent of those with bad karma.[17]

The Manussa Realm

The British Museum manuscript dedicates seven openings to level 27, the human world. Four of these illustrate the Himavanta region—one of animals, two of Buddhas on Gandhamādana mountain, one of the Buddha eating by the banks of Lake Anotatta surrounded by the five mountain ranges, and three schematic diagrams of cosmological structure and the sixteen *Mahājanapadās*. This material is located above the scenes of the Hell regions and presents the Manussa world in substantially different ways from either the Heavens or the Hells.

Scenes of the Himavanta forest start the Manussa sequence (Figure 2.6). Not only is there the requisite scene of elephants and *kinnara* and *kinnarī* (half-human, half-bird) in the forest on one fold, but the remaining two folds show Lake Anotatta and the surrounding rivers, mountains, and lakes, all duly named, with Buddhas scattered in the landscape. Gotama Buddha is shown seated under a tree with a *deva* paying homage, and *devas* also pay homage to *arahants* (enlightened beings) and *Pacceka* Buddhas seated in caves, standing under trees, and reclining on rocks in the landscape.

Although the scene of the Buddha eating by the banks of Lake Anotatta usually initiates the Manussa series, in the British Museum manuscript it

Figure 2.6. The Himavanta region. © The Trustees of the British Museum.

Figure 2.7. The Buddha by Lake Anotatta. © The Trustees of the British Museum.

concludes the sequence (Figure 2.7). Here, the stylized mountain ranges and lakes are labeled with their names, and an elephant and a *deva* pay homage to the Buddha who is portrayed sitting near the top of the page with his alms bowl. This scene represents the daily instances, during his three-month stay in Tāvatiṁsa Heaven to preach the *Abhidhamma* to his mother, when he would fly to the northern island of Uttarakuru to beg for sustenance and then travel to the shores of Lake Anotatta to eat.[18] In Buddhist texts, being able to access water from Lake Anotatta indicates extensive *iddhi* (supramundane) powers.[19] Together, the imagery thus proposes the supernatural powers of the Buddha, *arahants*, and *Pacceka* Buddhas, indicating that a result of enlightenment is the supernatural ability to access Himavanta and its benefits.

Unusually, the paintings of the Himavanta forest and the Buddhas on Gandhamādana Mountain over three folds are represented horizontally rather than vertically as the rest of the manuscript is. The horizontal organization suggests that the area was viewed as a single region plotted on the same stratigraphic line. The following schematic diagrams of the *Mahājanapadās* and Mount Meru and the four continents are arranged vertically, as is traditional, with the east at the top of the fold. The scene by the banks of Lake Anotatta is also organized vertically, as is made clear by the placement of the head of the Sīhamukha River, which is in the shape of a lion and springs from the lake's eastern side. Additionally, the Buddha is also seated at the top of the fold. This places the most sacred direction and the most sacred figure at the highest point on a page, reinforcing a vertically ethicized organization in keeping with the organization of the scenes of the Heavens and Hells. The arrangement of this latter image of Himavanta is vertical because it follows after the other schematic diagrams that are also oriented this way; perhaps if it had been attached to the other scenes of Himavanta, the orientation would have been horizontal.

After the initial Manussa images of Himavanta forest and the Buddhas on Gandhamādana Mountain is a schematic drawing of Mount Meru and the four surrounding continents (Figure 2.8) set within blue wavy lines indicating the surrounding oceans that cover the entire background space.[20] Meru is indicated by two concentric circular lines drawn in the center of the page, and each island is delineated by a double line in its specific shape and by its foremost tree, although all four are represented identically. Yet, within the quadrilateral form (described in the text as being in the shape of the front of a cart) indicating the southern island of Jambudīpa, there is also an image of the Buddha seated in *bhūmisparsa mudra*, the posture of enlightenment. While Buddhas may use a variety of places for meditation and other purposes, it is on Jambudīpa that enlightenment is possible, as this representation demonstrates. The location of the Buddha's great achievement thus becomes the important feature of this diagram, and the image shifts the emphasis from geography to the enlightenment and sacred biography by linking the cosmic map with the Buddha's awakening.

The two remaining schematic drawings are of the *Mahājanapadās*, the sixteen lands. One displays a central Bodhi tree surrounded by text mentioning the seven stations that the Buddha occupied during the seven weeks after his enlightenment, the names of the sixteen lands, and the boundaries

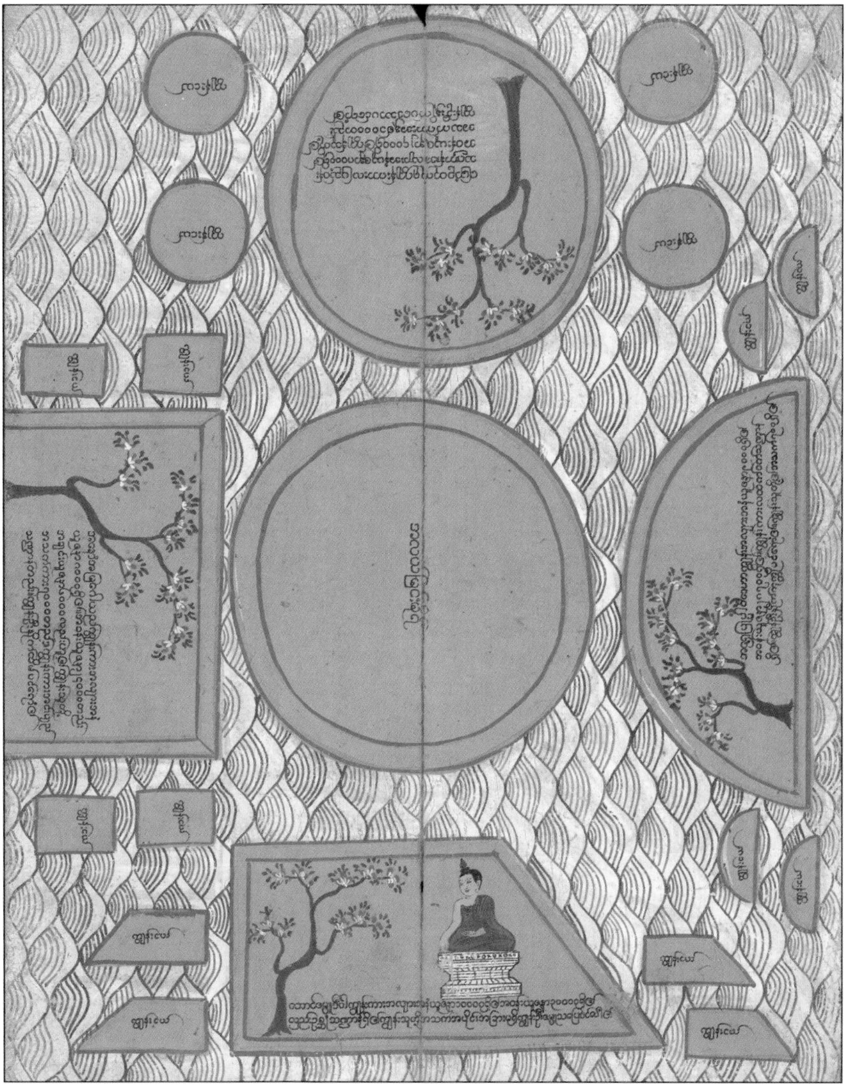

Figure 2.8. Mount Meru and the four surrounding continents. © The Trustees of the British Museum.

of Majjhimadesa, the area where it is possible to hear of the Buddha and his teachings (Figure 2.9). The second diagram portrays the Buddha seated in *bhūmisparsa mudra* in the center of concentric bands listing the sixteen lands (Figure 2.10). It is unusual to have two such diagrams, and if one is

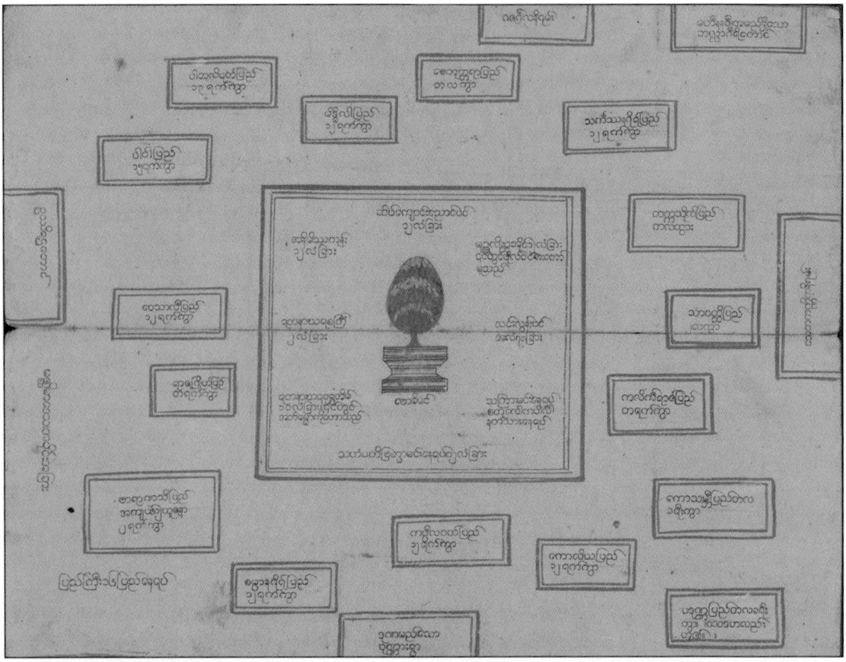

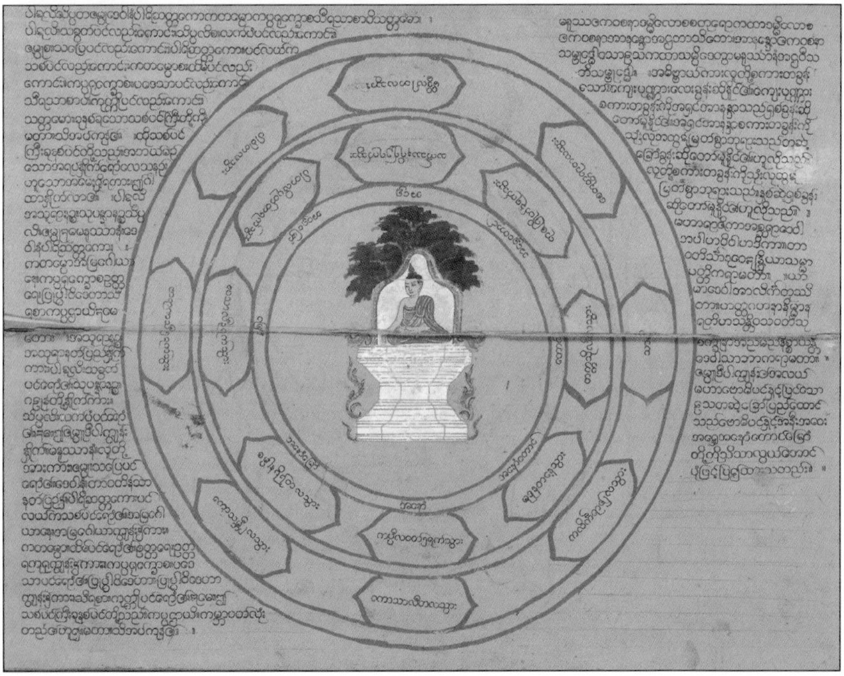

included, it is normally placed at the end of the manuscript rather than with the Manussa imagery. In the Manussa part of the British Museum's manuscript, the combination of *Mahājanapadā* diagrams and other material representing places associated with enlightenment focuses this section upon Gotama Buddha and the escape from the cycles of rebirth.

The Mahājanapadā *Diagrams*

The *Mahājanapadās* are places where the Buddha spent time during his final life (particularly during the rainy seasons) or previous lives, the countries of significant groups of people of the Buddha's lifetimes, or sites where the Buddha's relics are enshrined. There are various lists of the sixteen lands, but, unsurprisingly, the one used here is a Burmese version that can be found in other cosmology manuscripts. The Burmese *Mahājanapadās* include

Kapilavatthu: the Buddha's birth place

Sampannago (Campā): a city that the Buddha visited a number of times and a major trade center during his lifetime

Baranathi (B̆ārānasī): the sacred city near which the Buddha delivered his first sermon

Yasagyo (Rājagaha): the capital of King Bimbisara's kingdom, where the Buddha spent several rainy seasons, and the site of the first Buddhist council

---

Figure 2.9. The sixteen *Mahājanapadās,* diagram 1. © The Trustees of the British Museum.

Figure 2.10. The sixteen *Mahājanapadās,* diagram 2. © The Trustees of the British Museum.

Wethali (Vesālī): the capital of the Licchavi kingdom that was visited by the Buddha and where he stayed during two rainy seasons, the Licchavis enshrined his relics, and the second Buddhist council was held

Bawa (Pāvā): the city of the Mallas, who obtained relics of the Buddha after his cremation

Patalipouk (Pātaliputta): the capital of Magadha, one of the four chief kingdoms during the Buddha's life and where the third Buddhist council was held

Meiktila (Mithilā): the capital of Videha country, where the Buddha stayed and where relics were honored

Sedouktara (Jettutara): the city where Prince Vessantara was born

Thingathanago (Sankassa): the place where the Buddha descended to Earth after preaching to his mother in Tāvatiṁsa Heaven

Dekkatho (Takkasilā): a major center of Buddhist education

Thawaddy (Sāvatthi): one of six great cities during the Buddha's lifetime where he performed miracles and passed twenty-five rainy seasons at Jetuttara and Pubbarama monasteries

Kālinga: one of seven political divisions during the Buddha's lifetime and where a tooth relic was worshiped until it was taken to Sri Lanka[21]

Kosambī: a city of great importance during the Buddha's lifetime and the capital of the country ruled by King Udena whose stories feature in the *Dhammapada*

Koliyā: one of the major clans during the Buddha's lifetime and a region visited by the Buddha

Padunna: an unidentified site that in some Burmese manuscripts is called Mudu or Mudunna[22]

Of these sixteen lands, one relates to the Buddha's birth in his penultimate life, another (Takkasilā) was an important educational center, and the remainder, except for the unknown Padunna/Mudu, relate to events and people of his lifetime, as well as sites of his relics after the *Parinibbāna*. These locations are significant because they are associated with the Buddha, his biography, and the actions of devotees during and after his final life.[23]

In the diagrams, the *Mahājanapadā* sites are arranged according to direction around the Bodhi tree or the Buddha in *bhūmisparsa mudra*. In the first diagram, Mithilā and Jettutara are to the east; Sankassa and Takkasilā are to the southeast; Sāvatthi and Kālinga are to the south; Kosambī, Padunna, and Koliyā are in the southwest; Kapilavatthu and Campā are in the west; Bārānasī is in the northwest; Vesāli and Rājagaha are in the north; and Pāvā and Pātaliputta are in the northeast. The distances these occupy from the Bodhi tree, recorded in the diagram itself, vary from one day to two months, with seven sites recorded as being a twelve or fifteen days' journey away and five as one or two months' distant. The second diagram does not completely overlap with the first. The majority of the sixteen sites are one or two months away from the site of the Buddha's enlightenment, while the remainder are between one and five days distant. The twelve-day distance popular in the first diagram is absent here. The directions assigned to the locations are also slightly different. In the second diagram, Sankassa and Mithilā are in the east. Jettutara and Rājagaha are in the southeast; Takkasilā and Sāvatthi are south; Mudunna and Kālinga southwest; Kapilavatthu and Kosali (Koliyā) are in the west. Campā and Kosambī are to the northwest; Vesāli and Bārānasī are in the north, and Pāvā and Pātaliputta are northeast. Similar variations are found on other Burmese manuscripts,[24] which suggests that the actual distance and the preciseness of the direction are less important than the fact that the sites surround the place of enlightenment. The combination of location and distance measured in time posits the Buddha's dispensation throughout the space of Majjhimadesa and across temporal boundaries, translating it into a cosmically vast sacred geography.[25] As part of a cosmology manuscript, the diagrams of the *Mahājanapadās* illustrate time and space within the context of the universe's structure as a whole.[26] They therefore become a physical manifestation of how beings throughout their numerous rebirths interact with the Buddha and his relics in multiple locations.[27] These images thus have the effect of making the Buddha present over time, bringing him into the period of the manuscript's production and linking devotees in contemporary Burma with him.[28] And, through the placement of the imagery within the thirty-one realms of existence, the manuscript references the law of cause and effect governing rebirth and thereby indicates the good fortune of those who are part of the Buddha's community.

More subtly, the *Mahājanapadās* were also of consequence in the Burmese royal context.[29] During the establishment of royal cities—and this is particularly well documented in the Konbaung dynasty (1752–1885)—earth was gathered from four lakes, four islands, four hills, and sixteen former royal palaces to be mixed and plowed into the land of the new capital city. These are references to cosmological organizations—the four islands around Mount Meru, the four directions, and the sixteen *Mahājanapadās*, translated into royal spaces and thereby connecting the Buddha and kingship.[30] The incorporation of cosmologically and religiously representative material verified the new royal city as part of the Buddhist world system and the Buddha's dispensation, and the new site would hold great power because of these links. The capital thereby associated the Burmese king with cosmic structure, making him by implication a universal monarch.[31] The reference to the sixteen lands also mapped the Buddha's dispensation onto the kingdom as both a temporal (the Buddha's life) and a spatial (Majjhimadesa) phenomenon.

The two *Mahājanapadā* diagrams in the British Museum's manuscript present similar information in different ways. More frequently represented in cosmology manuscripts is the one with a Bodhi tree in the center surrounded by a frame within which is text that names the seven stations and their distances from the Bodhi tree, as well as mentioning the god Sakka, the four *Lokapālās*, and the *brahmā* Sahampati (see Figure 2.9). The outer section of the diagram has, in individual frames, the names of the sixteen *Mahājanapadās* and five locations, including Kajangala and Mahāsāla villages in the east, Salalavatī River in the southeast, Thūna Brahmin village in the west, and Usīraddhaja mountain range in the north, at the boundaries of Majjhimadesa. Beings are fortunate to be born here because outside this region, in Paccantajanapadā, there are few opportunities to engage with the Buddha. The inclusion of the edges of Majjhimadesa in this image implies the broad area where the Buddha's teachings are extant.

The diagram in Figure 2.9 has a number of unusual, but not unique, features. These include the substitution of the Bodhi tree for an image of the Buddha. The use of the tree firmly establishes the central point in time and space as the place and moment of the enlightenment in a way that an image of the Buddha seated in *bhūmisparsa mudra*, commonly used in a generic fashion in Burma, would not, and the *bodhimaṇḍa* as a victory ground is a suitable symbol of enlightenment.[32] The inclusion of the seven stations is another less common feature.[33] In the central box, the seven stations are arranged appropriately according to the directions with which they are associated. Each is indicated by text, rather than an image, that lists a distance in *lan*, which by implication is from the central Bodhi tree.[34] Additionally, in the lower right corner (the southwest), the text mentions that this is the place of the god of Tāvatiṁsa Heaven, Sakka, who donates medicinal fruit to the Buddha seven weeks after the enlightenment and the four *Lokapālās*, kings associated with the directions who live in Cātummahārājika Heaven. After the enlightenment, they gave the Buddha four alms bowls that he miraculously merged into one in order to accept the first lay offering of food from the merchant brothers Tapussa and Bhallika during the seventh week after enlightenment. No distance is recorded. Beneath the image of the Bodhi tree, to the west, the text states that Sahampati Brahmā, who urges the Buddha to preach, is twelve *lan* distant. The imagery of this central area thus focuses upon the events soon after the enlightenment. The circumscribed inner area of the *Mahājanapadā* diagram encapsulates the enlightenment and its immediate aftermath, including initial donations and the Buddha's decision to preach, connecting the directions and the area around the Bodhi tree with the establishment of fundamental activities—the commencement of the Buddha's dispensation and the establishment of giving and preaching as ritual practices. In the diagram, the sixteen *Mahājanapadās* form a concentric ring around the inner area, suggesting the expansion of the Buddha's dispensation and the activities suggested by the text across Majjhimadesa. The connections between the location of a site and an event that occurred at a specific locale (the enlightenment) are evidenced through the plotting of the places on the map and stating their distance from the Bodhi tree. In this diagram the viewer is made aware of the enlightenment, the Buddha meditating on the *Abhidhamma* and experiencing the bliss of emancipation, the establishment of giving as a significant activity, the decision to preach, and the physical extent of that preaching in the world of humans. The diagram is therefore a visual manifestation of the connection between space, time, the *Dhamma* (the Buddha's teachings), and sacred biography, and the map reveals the mutual dependence between sacred geography and sacred biography. Additionally, the schema emphasizes the main focus of Theravada Buddhist practice—the enlightenment. Everything about the Buddha—lives, relics, teachings—is posited in relationship to that event by emphasizing spatial and temporal relationships visually.

The second diagram of the *Mahājanapadās* in the British Museum manuscript presents the sites in two concentric circles around the Buddha seated in *bhumisparsa mudra* under the Bodhi tree (Figure 2.10). Like the first diagram, this one also shows the enlightenment as a central, defining moment. The use of two circles to enclose the sixteen lands is an unusual organization, but there are a number of other images in Burmese art and material culture of this type. Most obviously, the concentric, circular arrangement mirrors circumambulatory pathways around a central, sacred point—a mandala form with the sixteen lands marking the cardinal and intercardinal points. The imagery is carefully laid out with two of the sites at each of the eight points of the compass. The circle also represents one of the thirty-two marks of a great man, in this instance, the Buddhist Wheel of the Law representing the *Dhamma*.

Circular forms with twelve divisions are also associated with horoscopes and protective devices.[35] In Burma, there are protective devices that resemble the arrangement seen in the *Mahājanapadā* diagram in Figure 2.10. In wall paintings and on square pieces of cloth, such circular diagrams display a Buddha image in the center of two concentric rings. The inner ring is divided into twelve sections, nine of which contain a kneeling monk, usually particularly honored disciples, and the remaining three display the monk Shin Upagutta, the god Sakka, and the *Pañcavaggi*, the five monks who cared for the Buddha during his period of austerities and who became his first disciples. The outer ring also has twelve divisions that contain imagery associated with powerful beings, such as *nāgas*, hermits, monks, and the accoutrements of a universal monarch (an elephant and a horse together). According to the *Gazetteer of Upper Burma and the Shan States*, such illustrations painted on cloth were used by soldiers as turbans to protect them during battle.[36] Texts on some of these cloths confirm an efficacy in protecting the wearer from attacks by wild animals, sickness, fire, thieves, and so forth. There is also a connection with the zodiac in these diagrams, as indicated by the use of two sets of twelve compartments and the fact that some of the texts in the diagrams specifically link each of the personages represented with a zodiacal sign. A Thai manuscript at the British Library displays similar, concentric circular forms with powerful kings and animals from the annual zodiac within the twelve divisions per ring; so this diagram as a protective device is not specific to Burma and indicates a regional sharing.[37] The diagram in the British Museum cosmology manuscript places the Buddha's dispensation within cosmic space through the directional arrangement and indicates temporal progression through the use of the number twelve associated with the zodiac. These are then combined in a format associated with personal protection. Like the *parittas* of the twenty-eight Buddhas, the Eight Victories, and the Ten Great Jataka stories that enumerate beneficial names, titles, and events, the *Mahājanapadās* in this instance are presented as a protective list.

Conclusion

The British Museum cosmology manuscript largely follows a normative format for such objects. Standardization among the arts, particularly painting, has a long history in Burma, and is expressed by a commonality of layout and information combined with variations in small details. The similarity of Burmese cosmology manuscripts provides a specific image of cosmic order and indicates that a being's position within the cosmic schema relates to karmic achievements. The text also reinforces the concept of karma in relationship to consciousness and states of mind, discussing emotions like fear, anger, folly, and the comprehension of one's actions and the consequences thereof. The Heavens and Hells present a map of the universe as a metaphor for spiritual progress, providing insights into the rewards and punishments that befall inhabitants of the thirty-one realms of the cosmos in keeping with Buddhist practice. The imagery of the Manussa realm, with its schematic diagrams of the *Mahājanapadās* and Mount Meru and the four continents, and the lush imagery of Himavanta, takes a different tack to the other pictures. Since this region is the place where Buddhas arise, it is not surprising that the material depicted relates primarily to Buddhas, enlightenment, and the supramundane powers that result from such an attainment. The British Museum's cosmology manuscript thus visually represents cosmic order and reveals the Manussa realm as the place of enlightenment. As the manuscript would have been a religious donation, it participates in one of the dominant discourses of Buddhism—the merit path to felicitous rebirths and enlightenment—and thereby ensures the maintenance of the cosmic structure that it illustrates.

Notes

1 I would like to thank San San May for her invaluable assistance and Justin McDaniel for his patience and encouragement. John Okell kindly assisted me in figuring out how to understand a *nissaya* and helped translate some difficult parts of the text. Thanks are also due to Jana Igunma, Ashley Thompson, and William Pruitt.For specifics, see Frank E. Reynolds and Mani B. Reynolds, *Three Worlds According to King Ruang: A Thai Buddhist Cosmology* (Berkeley: Center for South and Southeast Asian Studies, University of California, 1982).

2 For further details, see Joseph E. Schwartzberg, "Cosmography in Southeast Asia," in *History of Cartography*, vol. 2, book 2, ed. J. B. Harley and David Woodward, 714–17 (Chicago: University of Chicago Press, 1994).

3 This format was also adopted in Thailand, but it is seen there at a much later date.

4 For discussions of materialization, see Elizabeth DeMarrais, "The Materialization of Culture," in *Rethinking Materiality: The Engagement of Mind with the Material*

World, ed. Elizabeth DeMarrais, Chris Gosden, and Colin Renfrew, 11–13 (Cambridge: McDonald Institute for Archaeological Research, 2004).

5 Juliane Schober, "Venerating the Buddha's Remains in Burma: From Solitary Practice to the Cultural Hegemony of Communities," *Journal of Burma Studies* 6 (2001): 125.

6 Peter Skilling, "Worship and Devotional Life: Buddhist Devotional Life in Southeast Asia," in *Encyclopedia of Religion*, vol. 14, 2nd ed., ed. Lindsay Jones, 9826–34 (Detroit: Thomson/Gale, 2005).

7 For arguments about the agency of objects, see Chris Gosden, "What Do Objects Want?" *Journal of Archaeological Method and Theory* 12, no. 3 (2005): 193–211; Alfred Gell, *Art and Agency: An Anthropological Theory* (Oxford: Clarendon, 1998); Howard Morphy, "Art as Action, Art as Evidence," in *The Oxford Handbook of Material Culture Studies*, ed. Dan Hicks and Mary C. Beaudry, 265–88 (Oxford: Oxford University Press, 2010); Dan Hicks, "The Material-Cultural Turn: Event and Effect," in ibid., 25–98.

8 Patricia Herbert, "Burmese Cosmological Manuscripts," in *Burma: Art and Archaeology*, ed. Alexandra Green and T. Richard Blurton, 77–97 (London: British Museum Press, 2002).

9 Ibid., 90.

10 Ibid. Richard Gombrich discusses how cosmological structure reifies spiritual progress. Gombrich, "Ancient Indian Cosmology," in *Ancient Cosmologies*, ed. C. Blacker and M. Loewe, 134 (London: Allen and Unwin, 1975).

11 See Michael Aung-Thwin, "Heaven, Earth, and the Supernatural World: Dimensions of the Exemplary Center in Burmese History," in *The City as a Sacred Center: Essays on Six Asian Contexts*, ed. Bardwell Smith and Holly Baker Reynolds, 94–95 (Leiden: Brill, 1987).

12 Herbert, "Burmese Cosmological Manuscripts" (note 8), 83.

13 Tusita is the Heaven where future Buddhas are born in their penultimate lives.

14 Herbert, "Burmese Cosmological Manuscripts" (note 8), 86.

15 Ibid. A manuscript in a private collection depicts a similar image with text stating that the woman deceived her husband and hence is being punished.

16 Ibid.

17 Gombrich, "Ancient Indian Cosmology" (note 10), 119.

18 G. P. Malalasekera, trans., *Dictionary of Pali Proper Names*, vol. 1 (Oxford: Pali Text Society, 1997), 98. See also Vin. i. 28 (Vinaya-Piṭaka) and DhA. iii. 222/3 (Dhammapada).

19 Ibid.

20 There is extensive literature on the importance of directions and directional orientation in Southeast Asia. See Schwartzberg, "Cosmography in Southeast Asia" (note 2), 738.

21 A country not on the list of sixteen *Janapadās* in the *Anguttara Nikāya* (A.i.213) but that is on an extended list of the *Niddesa* (CNid.ii.37). See Malalasekera, *Dictionary of Pali Proper Names* (note 18), online entry on Kālinga. http://www.palikanon.com/english/pali_names/ka/kaalinga.htm Accessed on 18 December 2014.

22 Kevin Trainor notes that one of the sixteen pilgrimage sites in Sri Lanka has also not yet been identified. Trainor, *Relics, Ritual, and Representation in Buddhism* (Cambridge: Cambridge University Press, 1997), 158.

23 For a discussion of the importance of place, see Richard A. O'Connor, "Transformations and Continuities: Sacralization, Place, and Memory in Contemporary Bangkok," in *Sacred Places and Modern Landscapes: Sacred Geography and Social-Religious Transformations in South and Southeast Asia*, ed. Ronald A. Lukens-Bull, 31 (Tempe: Monograph Series Press, Program for Southeast Asian Studies, Arizona State University, 2003).

24 Herbert, "Burmese Cosmological Manuscripts" (note 8), 92.

25 Juliane Schober, "Mapping the Sacred," in *Sacred Places and Modern Landscapes* (note 23), 11.

26 See I. W. Mabbett, "The Symbolism of Mount Meru," *History of Religions* 23, no. 1 (1983): 68.

27 Schober, "Mapping the Sacred" (note 25), 16.

28 Ibid, 8; Trainor, *Relics, Ritual, and Representation* (note 22), 188; Angela Chiu has written that the establishment of relics is multitemporal, demonstrating that past activity has significance for the present. Because of the relics' continuing significance, they ensure unity across time, linking past and present. Chiu, *The Social and Religious World of Northern Thai Buddha Images: Art, Lineage, Power and Place in Lan Na Monastic Chronicles* (Ph.D. dissertation, School of Oriental and African Studies, University of London, 2012), 124 and 125.

29 See Tun Aung Chain, "Prophecy and Planets: Forms of Legitimation of the Royal City in Myanmar," in *Myanmar Two Millennia*, part 3 (Yangon: Universities Historical Research Centre, 2000 [republished in *Selected Writings of Tun Aung Chain*, 2004]), 140–41, for information on how the royal cities physically represented time and space.

30 François Tainturier, *The Foundation of Mandalay by King Mindon* (Ph.D. dissertation, School of Oriental and African Studies, University of London, 2010), 40–41.

31 Ibid., 41; and F. K. Lehman, "Monasteries, Palaces, and Ambiguities: Burmese Sa-

cred and Secular Space," *Contributions to Indian Sociology* 21, no. 1 (1987): 176. The spire on buildings symbolically indicates secular and sacred power; the *cakkavattī* (universal monarch) represents the intersection of the sacred and secular domains, with the king as secular and the Buddha sacred.

32 *Bodhimaṇḍa* is of permanent significance because it is a victory ground (Burmese: *aung mye*). See Tun Aung Chain, "Prophecy and Planets" (note 29), 136; and Schober, "Mapping the Sacred" (note 25), 16. Burmese believe that localities must be strong to hold the Buddha's remains; in other words, they must be victory grounds.

33 See Catherine Raymond, "The Seven Weeks: A 19th-Century Burmese Palm-Leaf Manuscript," *Journal of Burma Studies* 14 (2010): 255–67.

34 One *lan* is equal to two yards.

35 Juliane Schober, "Burmese Horoscopes," *South East Asian Review* 5, no. 1 (1980): 52–54.

36 James G. Scott and J. P. Hardiman, *Gazetteer of Upper Burma and the Shan States*, vol. 2, pt. 1 (Rangoon: Superintendent of Government Printing, 1900–1901), 79–80, and fig. 1.

37 London, British Library, Add Ms 27370, folios 4 verso and 5 recto.

CHAPTER 3

Stories Steeped in Gold

Narrative Scenes of the Decorative Kammavācā Manuscripts of Burma

SINÉAD WARD

INSPIRED BY A MANUSCRIPT that formed part of the late Henry Ginsburg's collection (now part of the Asian Art Museum Collection in San Francisco), this essay will examine the introduction of narrative scenes in the decorative *Kammavācā* manuscripts of Burma (Figure 3.1). Such scenes are rare, appearing in only 10 of the 440 manuscripts surveyed to date. These manuscripts draw on themes that have a long and deep history in Southeast Asian art; however, their introduction marked a notable change from the centuries-old tradition of ornamentation. What caused this departure from traditional *Kammavācā* ornamentation to the introduction of narrative scenes? Unlike wall paintings or *parabaik* (paper manuscripts), the scenes depicted in the *Kammavācā* manuscript were not intended for a lay audience. What then was their purpose? Was the introduction of this type of illustration influenced by social change in late nineteenth- to early twentieth-century Burma? To understand the significance of these rare manuscripts we must first examine the traditional *Kammavācā* manuscript of Burma.

Kammavācā *Tradition*

The term *Kammavācā* refers to a collection of Pali texts for use in ceremonies by the Buddhist *Saṅgha* (monastic community). While the text of the *Kammavācā* is known throughout the Theravada world, it is only in Burma

Figure 3.1. Renunciation and departure scenes. Gift of Jared C. Ede, in memory of Emily Mead Baldwin, 2008.90. © Asian Art Museum, San Francisco. Used by permission.

that it takes the form of a decorative manuscript. The texts included are excerpts from the *Vinaya* and are arranged to form a coherent text, suitable for recitation in monastic ceremonies.

These manuscripts were commissioned and produced by the lay community for use by the monastic community. For the lay donor, the manuscripts constituted an important source of merit and were often commissioned to mark the ordination of a new monk. In Burma, the manuscripts are to this day produced in a lay workshop setting.

A variety of support materials were employed to produce these manuscripts, from palm leaf to ivory sheets. Lacquer was applied directly to the support to form the base to which the text and ornamentation would be applied. Owing to the layering technique used to apply the lacquer, the manuscripts could take months to produce. A decorative layer of gold, silver, or *mo-gyo* (a combination of gold and silver) leaf was applied to the lacquered support, using the same *shwe-zawa* (application of lacquer and decorative leaf) technique employed in the ornamentation of lacquered objects. The quality of decoration varies in direct relation to the means of the commissioner.

The text in the decorative *Kammavācā* manuscripts follows the same format as other palm-leaf manuscripts. However, the obverse of the first folio and reverse of the final folio are entirely dedicated to ornamentation (Figure 3.2). Text begins on the reverse of the first folio and ends on the obverse of the final folio. Large, decorative panels appear in the margins of both these folios, as well as on the obverse of the second and reverse of the penultimate folios. The remainder of the folios have a large text panel and are framed to the left and right by a small margin of ornamentation.

The decorative areas of the *Kammavācā* manuscript, as outlined above, were traditionally ornamented with motifs and later with figures unrelated to the text itself. The illustrations ranged from geometric and floral ornaments, to birds and mythical creatures, and later to flying *Devas* and *Kinnarī*, as well as the twenty-eight previous Buddhas in *Bhumisparsa Mudrā*. While such animals and figures were included in the works, they were not intended to depict a narrative scene; rather, they were framed by such devices as stepped squares, roundels, and intertwined knots. And it is the departure from such a rigid tradition that makes the introduction of

Figure 3.2. Traditional *Kammavācā* manuscript. Ms KV-187, Fragile Palm Leaves Collection, Bangkok. Reproduced by permission.

narrative scenes so remarkable. The narrative scenes that were introduced were confined to the areas of full ornamentation on the opening and closing folios; that is, they did not continue on to the large margin panels, which reverted to the traditional ornamentation.

Of the 440 *Kammavācā* manuscripts included in my survey to date, fewer than 1.5 percent (or five manuscripts) include ornamentation that would fall into the category of narrative scenes. This small minority of manuscripts is indeed quite exceptional, but how and why did this new trend emerge? All but one of the manuscripts demonstrate a masterful display of artistic skill and fine ornamentation. Are these works the product of a small number of craftspeople of exceptional skill? Was this trend inspired by a change in clientele or by the redeployment of craftspeople from other traditions? And where is evidence of the evolution from traditional ornamentation to the exceptional realization of narrative scenes to be found?

The manuscripts surveyed are purposely split between those that remain in Southeast Asian collections and those in European collections. This allows for an examination of manuscripts that left Burma during the colonial period, as well as those produced and collected after that time. My survey conducted over the past six years focuses on the manuscript leaves. Of the 440 manuscripts, 58 percent are from Asian collections, with 42 percent from European collections. The majority of manuscripts surveyed in Asia were acquired in the mid-1990s as they arrived on the local art markets. They are, therefore, of a considerably wider date range than those of the European collections, which made their way to Europe one hundred to two hundred years earlier. Unfortunately, not all the manuscripts surveyed in Asia were fully accessible, and many are recorded as they were displayed in museum cases: without accurate measurements, recording only what was visible at the time of viewing.

The Manuscripts Discussed

The manuscripts from the *Kammavācā* survey that will be discussed here are from the Fragile Palm Leaves Collection in Thailand, the Shwedagon Library in Burma, a private collection in Burma, and the British Library (BL). The Asian Art Museum (AAM) manuscript, formerly in the collection of Henry Ginsburg, will be included in the discussion but is not included in the manuscript database.[1]

The manuscripts listed below in their entirety are in gold leaf on a red lacquered ground. On one of the manuscripts from the Fragile Palm Leaves Collection (FPL KV-84), the ornamentation appears on the inside of the manuscript cover and not on the manuscript folios; however, it is included here because of its remarkable similarity to the narrative scenes in another manuscript.

The following manuscripts are discussed here:

London, British Library, Or 13896: metal support with six lines of *magyi-zi*[2] script.

Bangkok, Fragile Palm Leaves Collection, KV-84: cloth support with six lines of *magyi-zi* script. The narrative scenes appear on the reverse of the covers, not on the manuscript folios. An inscription on the cover names the illustrator as Maung Ba Nyo.

Yangon, Private Collection: cloth support with six lines of *magyi-zi* script. An inscription on the cover dates the manuscript to 1928 and identifies the donors as residents of Tharawaddy District in Bago Division. This is an unusual manuscript in that it contains three sets of full folio ornamentation (that is, six narrative panels in total).

Yangon, Shwedagon Library, 988: cloth support with six lines of *magyi-zi* script. An inscription on the cover dates the manuscript to 1928 CE (SL no. 988).

Bangkok, Fragile Palm Leaves Collection, KV-141: cloth support with six lines of *magyi-zi* script. A single narrative folio survives.

San Francisco, Asian Art Museum, 2008.90: cloth support with five lines of *magyi-zi* script. This manuscript was formerly in the collection of Henry Ginsburg.

Scenes Depicted

The narrative scenes depicted in these *Kammavācā* manuscripts are taken from three of the *Mahānipāta* (final ten) *Jātaka* stories and also from the life of the Buddha, reflecting a well-established narrative tradition in Burma. The *Mahānipāta Jātakas* represented here are the final three tales, the *Vidhūra, Mahā-Ummagga* and *Vessantara Jātakas*, popular themes in Southeast Asian art that illustrate the many perfections the Buddha-to-be achieved in his previous lives. The *Vessantara Jātaka*, one of the most popular *Jātakas*, is widely depicted in various art forms and is part of many rituals, from marriage ceremonies to the consecration of Buddha images.[3] It is thought by many Theravada Buddhists to be representative of the greatest perfection achievable. This is most likely owing to its position in the *Jātaka pali* as the final life before Siddhattha's life as the Buddha.[4]

The choice of scene depicted would have been heavily influenced by versions of the stories already in circulation, whether they were from oral, textual, or visual traditions. *Kammavācā* workshops were based around the royal courts in Amarapura and Mandalay during the Konbaung era, giving those craftspeople ample opportunities to explore new forms of visual and textual sources. No evidence exists of workshops outside the royal cities,

Figure 3.3. Sakka playing the harp, Life of the Buddha. © British Library Board, Or 13896.

but we can only presume that if they did exist (and research would indicate that there must have been regional *Kammavācā* workshops), they too would have been influenced by contact with other art forms.[5] The movement of people across Southeast Asia through trade, pilgrimage, religious studies, and warfare would have also led to a greater awareness of narrative traditions.[6]

A variety of textual sources were also an important influence on the origins of the narrative scenes. Alexandra Green, in her article "From Gold Leaf to Buddhist Hagiographies," discusses the significance of the *Pathamasambodhi* (and particularly the Tai *Khün* version of the text) on the illustration of certain narrative scenes from the life of the Buddha in Burmese art. One of these scenes, of Sakka playing the harp to the Buddha, is depicted in the British Library manuscript included here (Figure 3.3). The *Pathamasambodhi* text was known in oral tradition in Burma, but no written examples of the text have been found. This leads Green to conclude that the presence of such scenes in Burma was influenced by the oral traditions of Lan Na and the Shan states.[7] The *Kammavācā* craftspeople were from the lay community and were therefore unlikely to read the Pali *Jātaka* or life of the Buddha texts. It is therefore probable that they would have been influenced more by oral as opposed to textual narrative traditions. However the *Mala Lingara Wuthtu*, a Burmese prose version of the life of the Buddha, thought to be the source for painted *parabaik* versions of these stories, was in circulation by the early nineteenth century.[8] The remainder of the life of the Buddha scenes depicted in the *Kammavācās* discussed here (renunciation, departure, and tonsure scenes) are standard representations and could originate from any one of the aforementioned sources.

Edward Cowell's translations from the Pali texts form the basis of the *Jātaka* stories discussed below.[9] However, as Catherine Raymond has pointed out, it is important to look for subtle regional differences in the stories, which may not always be obvious when working from Cowell's translations.[10] While choice of scenes may be standardized across Theravada countries, localizations appear in the form of additional or unusual representations that may not fit those of the original textual source.[11]

Vidhūrapandita Jātaka: *The Virtue of Perfect Truth* [JA 545]

Scenes from the *Vidhūrapandita Jātaka* can be found in Yangon, Private Collection, (Private Collection) ms. 1 (two folios), Fragile Palm Leaves Collection (FPL) KV-84 (two covers), and FPL KV-141 (one folio).[12]

The *Vidhūra Jātaka* is based on the tale of Vidhūrapandita, the Buddha-to-be, who was at the time of the story, a minister to King Dhanañjaya. He was renowned for his discourse and intelligence. The wife of the Naga king Varuna, having heard of his great teachings, wished to hear Vidhūra speak. She led her husband to believe that she would die if she did not have the minister's heart. The Naga king enlisted the help of his daughter, who promised to marry the yakka (ogre) general Punnaka, if he would fulfill her mother's wishes. The ogre Punnaka, aware that King Dhanañjaya would not easily relinquish his minister, decided to play on the king's weakness for gambling. He traveled to the Kuru kingdom disguised as a prince and defeated the king in a game of dice, claiming the minister as his prize.

All three of the manuscripts depict this dice-throwing scene in a similar fashion (Figure 3.4). On the left of the panel, King Dhanañjaya and Punnaka are seated in a royal pavilion, with one or more attendants (including in two of the manuscripts, the minister Vidhūra). Punnaka, disguised as a human, is identified by the ogre mask that he wears on his head. All three scenes are framed to the left by a tree (a palm tree in two instances) and tiered roof that are visible over the palace wall. A figure in royal clothing floats above the wall viewing the scene. This celestial being is most probably King Dhanañjaya's mother from a previous life, who acting as the king's guardian, always ensured he won the game. In order to defeat the king using his own powers, Punnaka had first to frighten the goddess and make her flee, leaving the king without a guardian. To the right of this scene, separated from the dice playing by trees and a large gateway, Punnaka appears with his horse. In Yangon, Private Collection s.1 and Fragile Palm Leaves Collection, (FPL) KV-141, Punnaka converses with the minister before tying him to the tail of the horse. The third representation, in con-

Figure 3.4. King Dhanañjaya and Punnaka play dice, and Punnaka converses with Vidhūra, *Vidhūrapandita Jātaka*. Ms KV-141, Fragile Palm Leaves Collection, Bangkok. Reproduced by permission.

Figure 3.5. Vidhūra flies through the air, and Vidhūra held aloft by Punnaka, *Vidhūrapandita Jātaka*. Private collection. Reproduced by permission.

trast, has the general on horseback in what appears to be his arrival at the court and not his departure.

The second panel of both manuscripts tells the remainder of the tale in four episodes, each separated by trees. In the first episode, Punnaka, having revealed himself as an ogre, leaves for the Black Mountain, with the minister tied to the tail of Punnaka's horse as he flies through the air (Figure 3.5). In the second episode, Punnaka attempts to kill Vidhūra by flinging the minister around over his head to obtain his heart for the queen. Vidhūra remains unfazed by this treatment and simply asks why this Yakka wishes to kill him. Having realized that the Naga king must have been deceived by the queen, Vidhūra requests that Punnaka place him on the top of the mountain so that he can teach him the Law of Good Men. The third episode on both panels depicts this scene, with the Yakka general kneeling at the feet of the minister, who appears seated on a pedestal (FPL KV-84) or on the mountain (Private Collection ms.1) as his horse stands nearby. The final scene used to illustrate the *Vidhūra Jātaka* is similar to the previous section, but here Vidhūra expounds the Law of Good Men to the Naga king, who kneels at his feet, identified by his Naga crown.

The narrative works well in this context, successfully illustrating the complete tale in a very compact format. It also includes two iconic scenes (Vidhūra flying through the air behind the horse, and Vidhūra held aloft over Punnaka's head) that are often used to represent this *Jātaka* .

Mahā-Ummagga Jātaka: *The Virtue of Perfect Knowledge [JA 546]* (Mahosadha Jātaka)

The *Mahā-Ummagga Jātaka* illustrates the perfection of knowledge and is depicted on two folios of both Yangon, Private Collection ms. 1 and Shwedagon Library (SL) no. 988. The manuscripts are almost identical in their representation of the tale, yet without a short caption that appears on two of the folios, the story would be difficult to identify. The caption identifies the episode of Mahosadha fighting Kevatta, a central part of the *Mahā-Ummagga Jātaka*.

At the time of this tale, the Buddha-to-be was Mahosadha, a wise sage in the court of King Videha, who was determined to protect the city of Mithilā. Upon receiving word that King Culani Brahmadatta of Uttarapancala was taking control of all the kingdoms of India, Mahosadha prepared to outwit the king and his advisor Kevatta. As Kevatta's many attempts to take the city failed, he decided to challenge Mahosadha to a Battle of Law in the belief that this would allow him to outwit the sage. But Mahosadha proved too wise to be outwitted by Kevatta. On the prescribed day, each traveled to the meeting spot accompanied by his warriors. The princes from the previously conquered Indian kingdoms were also present. While handing a precious gem to Kevatta as a gift, Mahosadha caused it to fall to the ground, resting at his feet. As Kevatta bent down to pick up the gem, the sage restrained him with his hand, preventing him from rising. Kevatta's supporters who had witnessed the scene believed that Kevatta was bowing to Mahosadha in deference, and they fled, afraid that they too would be conquered by King Videha.

Figure 3.6. Military parade, *Mahā-Ummagga Jātaka*. Private collection. Reproduced by permission.

Figure 3.7. Mahosadha outwits Kevatta, *Mahā-Ummagga Jātaka*. Top: Shwedagon Library, Yangon. Bottom: Private collection. Reproduced by permission.

It would appear that the first of the two Mahosadha episodes in these *Kammavācā* manuscripts depicts Mahosadha as he prepares to meet Kevatta, leaving the city on horseback (not the chariot mentioned in the *Jātaka* tale) accompanied by his warriors in a military parade (Figure 3.6). However, without any accompanying text, an exact identification is impossible to determine. The second episode focuses on the Battle of Law (Figure 3.7). Mahosadha and Kevatta are sheltered by their respective royal umbrellas, their supporters at their side. As Kevatta bends down to pick up the gem, Mahosadha places a hand on his head. To the left Mahosadha's supporters appear to rejoice, while on the right Kevatta's supporters turn and prepare to flee.

This is a somewhat truncated version of the *Mahā-Ummagga Jātaka*. The many trials to establish the extent of Mahosadha's wisdom are excluded, as is the final tunnel episode, in which Mahosadha finally overcomes King Brahmadatta. What is most striking here is the similarity of the illustrations in both manuscripts, which will be discussed later.

Vessantara Jātaka: *The Virtue of Perfect Generosity [JA 547]*

Of the manuscripts discussed here, only Private Collection ms. 1 includes scenes from the *Vessantara Jātaka*. Depicted over two folios, the *Vessantara Jātaka* illustrates the Buddha-to-be's perfect generosity, where he reaches the summit of perfection through his extraordinary giving. The Buddha-to-be was Prince Vessantara, who from a young age gave away all the goods he possessed. Nevertheless, Vessantara knew that to attain perfection, he would have to give something from within—something of his own.

Having been banished from his kingdom for giving away the white elephant and bringing on a drought, Vessantara travels to the forest with his wife and two children to live the life of a hermit. According to the *Jātaka* tale, they are drawn by four horses, and along the way they are approached by four Brahmin who request the horses from Vessantara.[13] In this manuscript, the family is seen leaving the palace walls in the horse-drawn chariot; however, the scene includes only two horses and two Brahmin (Figure 3.8). Meanwhile, the Brahmin Jujaka, spurred on by his wife, decides to visit Vessantara and ask for his two children as slaves. The second section of the first leaf depicts Jujaka's journey through the forest as he encounters an old hermit in rags whom he asks for directions.

Having obtained directions, Jujaka travels on to the hermitage, as depicted in the third scene on this first folio (Figure 3.9). The story continues that while Vessantara's wife, Maddi, is out gathering food, Jujaka

Figure 3.8. Vessantara travels to the forest; Jujaka meets the hermit, *Vessantara Jātaka*. Private collection. Reproduced by permission.

approaches Vessantara and requests that he take the children as slaves. Vessantara, in the knowledge that such an act would fulfill his need to give something of himself, thus attaining perfection through giving, agrees to give the children to Jujaka, an agreement signified by Vessantara pouring water over Jujaka's hands.[14] He later requests that Jujaka take the children to his father the king, who will reward him for their safe return, but Jujaka refuses and takes the children away with him as slaves. Jujaka is identified in these scenes by his particular headdress and traveling bag, which he carries across his back. The seated Vessantara holds prayer beads as he speaks with Jujaka. He has shed the royal clothing that he wore as he left the palace. However, he does not wear the animal skins that were often favored in earlier representations of the scene. In this instance, his clothing appears to be that of a monk, evidenced by the section of cloth draped over his left shoulder, beneath which lies a long-sleeved garment indicated by bands visible at the neck and wrists.[15] The children remain in royal attire throughout as opposed to the more common images where they are dressed in rags and skins.

Figure 3.9. Vessantara agrees to give his children to Jujaka, *Vessantara Jātaka*. Private collection. Reproduced by permission.

The second folio takes up the plight of the children who, according to the *Jātaka,* are bound up with a piece of creeper and driven on like cattle. Unbeknownst to Jujaka, *Devas* (divine beings) intervene to lead all three to the palace walls. The story now moves to the interior of the palace, where the children are reunited with their grandfather the king. The children, now slaves, must be bought back by the king. Having sold the children, Jujaka proceeds to eat himself to death, a scene illustrated in the far left of the folio, where the Brahmin sits alone with his eating pot in front of him.

This is the final episode from the *Vessantara Jātaka* in this manuscript. We do not observe the usual scenes of the children hiding in the lily pond, Maddi held back in the forest by wild beasts, the children comforted at night by the gods, the king traveling on elephant back to find his son, or Vessantara and Maddi reunited with their children. Owing to the considerable space restrictions, the episodes depicted are intended to hint at rather than to tell the story. Raymond draws on post–seventeenth century temple paintings as evidence that the Burmese artists had little interest in the final scenes from the *Vessantara Jātaka,* focusing instead on Vessantara's gift of his children.[16]

Life of the Buddha

The life of the Buddha scenes in both the Asian Art Museum (AAM) and British Library (BL) manuscripts take place before the Buddha's enlightenment. AAM 2008.90 is exceptionally well illustrated, certainly superior to almost all the other manuscripts discussed here. It succeeds in covering many scenes within the two available decorative panels, while the BL manuscript includes only one scene per panel.

AAM 2008.90 begins its story, set within the palace walls, with the renunciation scene. Prince Siddhattha, having already witnessed the four signs, decides to leave his family and seek *Nibbāna.* In this scene he visits the bedside of his wife and newborn son as they sleep to say goodbye (see Figure 3.1). Siddhattha's attendant Channa waits beside him. Separated by a tree and royal gate or doorway, Siddhattha's departure is depicted in the second scene. As Siddhattha leaves on horseback, Channa holds on to the horse's tail, while the sound of the hooves is silenced by the gods. Still within the palace grounds (unusual for this scene), the prince is confronted by Mara with an outstretched hand in an unsuccessful attempt to prevent Siddhattha from embarking on his path to enlightenment.

The Bodhisatta's tonsure scene is depicted in both AAM 2008.90 and BL Or 13896. In the AAM manuscript, this takes place in an idyllic forest setting where wild deer roam and fish are seen in the river and birds in the flowering trees (Figure 3.10). The seated prince, with two *Devas* to his right and Sakka and Ghaṭīikāra to his left, cuts off his long topknot. As he throws a section of hair into the air, it is caught by Sakka in a flower-shaped casket, a sign that the prince will become a Buddha. Meanwhile, Ghaṭīikāra kneels at the prince's feet, offering the alms bowl and fan of a monk. A less well-executed depiction of this episode may be found in the BL manuscript, taking the full width of the panel to illustrate this one scene (Figure 3.11). Here the Bodhisatta, seated by a river with lilies and ducks, cuts his long hair with his sword, while to his left Sakka holds out the casket to catch the hair. To the right of the scene appear two gods bearing the robes and alms bowl required by a monk. Two unusual characters are shown to the left of the seated Bodhisatta in the BL manuscript. Their dress and beards fit the appearance of hermits, and they appear to be carrying a pineapple and durian fruit. Perhaps this is a reference to the Bodhisatta's years spent as a wandering ascetic, eating only fruit and leaves in his search for enlightenment. Or it may refer to the hermits Alara and Uddaka, who were his teachers in the Uruvela forest. The depiction is certainly unclear, and the unidentified characters are unusual in this context.

To return to AAM 2008.90, the final scenes take place while still in the forest setting. The Bodhisatta dismisses his attendant Channa and his horse Kanthaka, who reluctantly leave him. In a striking depiction, Kanthaka dies of a broken heart, having no desire to leave the Bodhisatta (see Figure 3.10). With legs upturned and veins visible, the horse appears to hover in midair as Channa weeps behind him. The illustrations conclude with Channa traveling back to the palace on his own, carrying the saddle and bridle of the dead horse.

Figure 3.10. Tonsure scene and death of Kanthaka, Life of the Buddha. Gift of Jared C. Ede, in memory of Emily Mead Baldwin, 2008.90. © Asian Art Museum, San Francisco. Used by permission.

Figure 3.11. Tonsure scene, Life of the Buddha. © British Library Board, Or 13896.

The second folio from BL Or 13896 depicts an episode from the life of the Buddha that is recounted in the *Pathamasambodhi*.[17] To the left of the scene, Sakka plays a harp to the Bodhisatta, while three floating *Devas* appear to the right (see Figure 3.3). According to the *Pathamasambodhi*, Sakka played the three-stringed harp to the Bodhisatta in order to encourage him to abandon his austere lifestyle and find the path to enlightenment through meditation. Here, and in many Burmese versions of this scene, the many-stringed Burmese *saun* is substituted for the three-stringed harp.[18] The Bodhisatta is portrayed standing with his alms bowl by his side, an unusual position for this scene, where a seated or reclining position is the norm.[19]

Style and Dating

As stated previously, the episodes selected to illustrate these manuscripts may have been influenced by contemporaneous art forms. They closely follow the standard depictions that appear in many wall paintings. While the scenes chosen may simply reflect the most important events in the tales, two of the stories are missing their conclusion. In the *Vessantara Jātaka,* we do not see the children reunited with their parents. Similarly, Mahosadha's final triumph over King Brahmadatta is not depicted. The artists were working within a considerably restricted panel. With about 10 × 55 centimeters available for narrative scenes, such limitations might have led the artists to select easily recognizable episodes from existing examples in other media at the expense of the concluding scenes.

Various art forms, such as wall paintings, *parabaik* illustrations, and painted textiles, may also have influenced the direction that the visual narrative takes in these manuscripts. Two directional forms, the narrative and the visual movement, are at play in many of the scenes. Depending on the manuscript and folios, the narrative moves from right to left, or left to right, in continuous mode. Of the manuscripts detailed here, the two *Mahosadha* folio pairs (SL no. 988 and Private Collection ms. 1), as well as the British Library life of the Buddha folio pairs, illustrate one narrative episode per folio, preventing any analyses of the narrative direction. With the exception of the *Vidhūra* covers in FPL KV-84 (Figure 3.12) and the *Vessantara* folio pairs in Private Collection ms. 1, all other folios have a narrative moving from left to right. What we see in the case of these two exceptions, however, is movement in one direction in the first panel (that is, left to right), followed by the opposite movement in the second panel (that is, right to left). These anomalies are of interest for two reasons. First, while the typical direction of illustrations within the manuscript follows that of the text, with a left-to-right and top-to-bottom reading, the flow of the illustrations on the cover appears to change direction to allow for a continuous reading of the narrative while the manuscript is closed. The bundle of leaves (or in the case of FPL KV-84, the covers) may be turned to allow the viewer to experience the scenes in a clockwise (Private Collection ms. 1, *Vessantara* panels, see Figures 3.8 and 3.9) or counterclockwise (FPL KV-84, see Figure 3.12) motion. Second, the directional anomaly in FPL KV-84 is not repeated in the *Vidhūra* panels in Private Collection ms. 1, but the illustrations are otherwise closely related.[20] The *Vidhūra* scene on the covers of FPL KV-84 depicts the ogre arriving at the palace; Private Collection ms. 1 depicts the ogre leaving the palace. Was this the original layout in another medium? Or was the original layout adjusted to allow the two *Vidhūra* manuscript covers to be viewed as a continuous pair? The direction of the scenes may have been copied directly from murals or another medium, which was influenced by physical layout and movement within a sacred space.[21] When transferred to these manuscripts, however, the original context no longer applies, causing a right-to-left movement to appear unusual.

In addition to the narrative direction, there is a separate visual movement at play in these illustrated folios that may follow the same direction as the narrative or converge or diverge in the center of the panel. For example, two of the *Vidhūra* folios (Private Collection ms. 1 and FPL KV-141) illustrating the dice-throwing episode show movement toward the center of the panel. This is produced by the horse leading into the scene on the right, with the celestial figure to the left, gazing over the game of dice (see Figure 3.4). AAM 2008.90 makes use of diverging visual movement while illustrating the renunciation and departure episodes, with both scenes moving from the center toward the opposite edges of the panel (see Figure 3.1). AAM 2008.90's tonsure panel employs this visual movement to unite the separate scenes. Before the death of Kanthaka, all eyes rest on the Bodhisatta, with

***Figure* 3.12.** *Vidhūra Jātaka*, inside covers of Ms KV-84, Fragile Palm Leaves Collection, Bangkok. Reproduced by permission.

the various scenes being connected by the backward glance of both Channa and the horse (see Figure 3.10).

Scattered throughout these narrative illustrations we find caption boxes to allow for the addition of explanatory text. Whereas on wall paintings and *parabaik,* such captions appear below the scene, the *Kammavācā* producers were working with a significantly more restricted space. Caption boxes are included within the scene or disguised in the roof of a pavilion or the base of a throne. The AAM manuscript artist has provided caption space below the sleeping mother and child, but, unusually, it remains empty. While this renunciation scene requires little explanation, the *Mahā-Ummagga Jātaka* is difficult to identify from the illustrations alone and relies on accompanying captions to explain the scene (see Figure 3.7). The standard *Kammavācā* framing bands of beads, zigzags, and twisted ribbons are still employed to frame the narrative panels on BL Or 13896, AAM 2008.90, and FPL KV-141. The remaining three manuscripts contain panels framed by the mythical *to-naya* (snake-like dragon) within scrolling vines. Within all the panels, both natural and architectural features are employed to mark the spatial-temporal divide between narrative episodes. Such devices are commonly utilized in wall paintings.[22] Palace walls, gateways, and the trunks of trees appear often, as both a framing and dividing device. The *Kammavācā* artists have clearly drawn from many of the standard artistic conventions of their contemporaries and adapted them to suit the requirements of these miniature illustrations.

The dating of the manuscripts is not easily determined, as very few of the *Kammavācā* manuscripts contain information relating to their production. Such information was usually recorded on the accompanying *sa si gyo* (woven wrapping ribbons), many of which have been misplaced over time. The practice of including production and donation information on the cover or in the manuscript folios appears to be a component of a more recent style of *Kammavācā* production (late nineteenth century onward), as is the use of a larger support and a round, instead of *magyi-zi,* script. However, the manuscripts discussed here still retain the *magyi-zi* script, which may place them toward the cusp of these developments.

Private Collection ms. 1 and SL no. 988, which contain almost identical illustrations of the *Mahosadha Jātaka,* are accompanied by covers bearing inscriptions that date them to 1928 CE (1290 Myanmar era).[23] Although such wooden manuscript covers were often misplaced and later united with the wrong bundle of folios, these particular covers appear to be the originals. Both manuscripts are on a cloth support, with six lines of *magyi-zi* script per side. Private Collection ms. 1 also contains illustrations from the *Vidhūra* and *Vessantara Jātakas*, with the *Vidhūra Jātaka* scene bearing a remarkable similarity in terms of the scene layout and ornamentation to its equivalent in FPL KV-84 and to a lesser extent the illustrated folio in FPL KV-141. While FPL KV-84 is once again a cloth-support manuscript with six lines of *magyi-zi* script, the illustrations appear on the wooden covers. FPL KV-141 is also a cloth-support manuscript with six lines of *magyi-zi* script, although it is slightly smaller than those previously cited and is incomplete, with the notable absence of one of the fully ornamented folios. There are striking similarities in terms of style of drawing and layout of scenes across all four manuscripts. Combined with the similarity in structure, it is therefore highly plausible that they were produced by the same workshop if not by the same hand.

While the arrangement and scene selection of FPL KV-141 is similar to that of FPL KV-84 and Private Collection ms. 1, stylistically it is much closer to AAM 2008.90. The panel borders are identical, and their treatment of trees is very distinctive and different from those in other illustrations (see Figures 3.1, 3.4, 3.10 and 3.11). In addition, both carry similar brick constructions to the left of the scenes (the renunciation scene in the case of AAM 2008.90). FPL KV-141 demonstrates a good understanding of perspective, with the castle parapet in the center of the scene retreating into the distance, a skill not yet mastered in AAM 2008.90. There are also similar caption areas within the illustrations in both manuscripts; however, the AAM box (beneath the sleeping wife of the Buddha and child) is blank. In the AAM 2008.90 tonsure scene, all available space is filled with the scrolling flowers, leaves, and little parrots commonly found in many *Kammavācā* manuscripts; fish, monkeys, and deer, less familiar inclusions in traditional *Kammavācā* ornamentation, also make an appearance. In comparison, the departure of the Buddha in AAM 2008.90 is a far less cluttered scene. FPL KV-141 is more selective in its choice of additional features. While it includes trees with hanging leaves or fruit, it omits the profusion of birds and animals. Although the manuscripts may be considered stylistically similar, the skill evidenced in the illustrations differs. Notwithstanding the difficulties with perspective, AAM 2008.90 shows a steadier and more skilled hand. The illustrations in FPL KV-141 appear more free-flowing and display a greater understanding of depth and arrangement. One wonders whether these manuscripts were produced by artists familiar with each other's work.

With regard to the dating of these manuscripts, the layout of FPL KV-141 is quite similar to that of Private Collection ms. 1, which, if its covers are reliable, dates to 1928. AAM 2008.90 is accompanied by an inscription on its cover dated 1938, but the inscription may not be the date of the original work and may in fact be a later addition; it seems to have been added after the ornamentation. Recycling of both manuscript leaves and covers was not unknown in relation to *Kammavācā* manuscripts, both historically and to the present day. Evidence exists of a reused cover in the Fragile Palm Leaves Collection (the original gold-leaf ornamentation is visible where a section of *thayo* paste[24] and cut-glass stones have come away), and even today it is not unusual to find a secondhand *Kammavācā* leaf cleared of text and rewritten.[25] There is also the likelihood that the covers were not the originals. The majority of manuscripts surveyed are accompanied by generic covers or covers too large or too small for the manuscript, which suggests that they are regularly misplaced or interchanged. AAM 2008.90, therefore, would require closer inspection to determine whether these were in fact the original covers. Without such information, it is difficult to determine the date of the covers, the date of the manuscript, and whether the accompanying dedication was rewritten at a later date.

BL Or 13896, in contrast to the cloth manuscripts referred to above, is on a metal support, with six lines of *magyi-zi* script and two fully illustrated folios containing scenes from the life of the Buddha. Both the choice of scene and the rushed sketchy drawing are quite different from those in the other manuscripts discussed. The only similarity between this manuscript and the AAM manuscript is the overall theme of the life of the Buddha. Their style and execution of drawing bear no relation to each other, nor does the understanding and utilization of available space. In fact, the style of drawing displayed in the British Library folios has more in common with the traditionally ornamented *Kammavācā* manuscript than with these narrative scenes. Perhaps the British Library manuscript was produced for a moderately wealthy patron or is the product of a lesser workshop, endeavoring to replicate the narrative style discussed.

While the dating of the manuscripts is rarely exact, I would feel confident that based on the number of lines, the size, the material, and the break from traditional ornamentation, all of these manuscripts may be dated somewhere between the late nineteenth and early twentieth centuries. This is further evidenced by the inclusion of one British Library manuscript with narrative panels, which was not acquired by the India Office Library during the colonial era but rather purchased at Sotheby's in 1979.

Purpose of the Narratives

The purpose of narrative scenes in the *Kammavācā* manuscripts is unclear. They appear in a manuscript in which traditionally there was no connection between the written text and the decorative motifs. Yet the narrative themes selected for these manuscripts are particularly fitting. The scenes depicted in both the AAM and BL manuscripts refer to the Buddha's life pre-enlightenment. Both the tonsure and departure scenes were suitable references for the monks about to be ordained. However, if such works were commissioned for the ordination of a son, they may have held more significance for the donor family, marking their sons' departure from the family unit to follow the path of the Buddha.

In this context, it is unlikely that the scenes were intended for the purpose of a didactic function or for use as narrative prompts. Naomi Appleton, in her analysis of the purpose of the *Jātaka* tales, notes Brown's critique of Dehejia's assumption that all such images have a narrative function.[26] As Brown and Appleton point out, the scenes depicted in wall paintings are often inaccessible to viewers because of their location. The illustrations on manuscripts such as the *Kammavācā* are equally inaccessible. Use of the manuscript would have been restricted to one reader at a time while a ceremony was being conducted. As the folios are turned horizontally to read the text, the illustrations would appear upside down to anyone facing the reader. It is also unlikely that any monks present during a ceremony would be contemplating the illustrations. Their focus would have been on the ceremony being conducted. When not in use, the manuscript would be wrapped and stored, not put on display. Perhaps this explains why narrative scenes were not added to the manuscript for many centuries, and ornamentation remained just that: ornamentation, with no other purpose than to decorate the manuscript.

The removal of a narrative function allows for a closer examination of other possible factors at play in the development of these scenes. Both the *Jātaka* and life of the Buddha episodes might have been included as symbolic references, representing what a newly ordained monk may achieve if he follows the Buddha's teachings. The *Jātaka* tales not only indicate the many perfections necessary to advance toward *Nibbāna*; they also symbolize, through the historical character of the Buddha, an achievable means to enlightenment.[27] Can the same also be said of the life of the Buddha scenes? Renunciation, departure, and tonsure scenes would have held obvious significance for those joining the *Saṅgha*, while the inclusion of Sakka's harp playing acts as a reminder of the importance of meditation in reaching *Nibbāna*. Such references are not unique to these manuscripts; Alexandra Green identifies a similar emphasis on the path to enlightenment in wall paintings.[28]

For the donor, too, the emphasis was on attaining enlightenment, but the journey was recognized as a significantly longer one. *Kammavācā* manuscripts have always constituted an excellent source of merit. As a compilation of ceremonial texts, they are not only an important record of Buddhist literature but also essential for the functioning of a monastery. With the addition of narrative scenes from the previous lives of the Buddha, the manuscripts may have acquired an additional level of merit provision. Illustrations of the *Vessantara Jātaka* in many media were for many centuries produced as a source of merit. Their use in conjunction with such a significant text would surely have increased the *kamma* accumulated. Alexandra Green views the emphasis in wall paintings on the status of royal figures, their luxury and power, as a reminder to the viewer of the status that can be attained by accumulating such stores of good *kamma*. The offering of a *Kammavācā* manuscript was therefore an advancement toward a better rebirth and, through it, a speedier path to enlightenment. It is interesting to note that Green sees the twenty-eight Buddhas as a reminder that *Nibbāna* is the final goal, not the better rebirth.[29] I have noticed in the *Kammavācā* survey a departure from the depiction of the twenty-eight Buddhas. This may reflect a greater concern in the mind-set of the donors with the accumulation of merit rather than the final goal of *Nibbāna*. Perhaps the twenty-eight Buddhas *Kammavācās* had been commissioned by members of the royal courts, who had no need to reflect on their own status, instead focusing on that which was left to attain.

Appleton sees the *Jātakas* as a bridge between the Buddha and those who follow him.[30] Perhaps the inclusions of both the *Jātakas* and the life of the Buddha episodes are intended to encourage the manuscript donor and the monks who use those manuscripts to attain an achievable level of perfection.

Influence of Social Change

While the ornamentation may have changed, the overall function of the manuscripts examined here has not. They are still produced for donation, still generate merit, and mark a life passage for a novice monk. They are still produced by the same workshops, with the same degree of skill, based on the wealth of the commissioner. So what might have prompted the change from traditional ornamentation? The late nineteenth and early twentieth centuries were a time of great social change in Burma due to the end of the royal courts and the introduction of British rule. Unlike paintings and other art forms that were suitable for purchase by foreigners, the continuation of the *Kammavācā* tradition lay firmly in the patronage of the Burmese public. It is unlikely that these manuscripts were produced for a non-Buddhist.

The break from traditional ornamentation may be the result of much greater changes in Burmese society. It is possible that the breakup of the royal courts led to the introduction of a much wider clientele to the *Kammavācā* industry. New patrons may have been interested in commissioning narrative scenes familiar to them. The *Kammavācā* industry, although based around the royal courts for quite some time, survived the annexation of Upper Burma in 1885. Of the Western collections surveyed, the majority of manuscripts were taken from Burma following the British annexation. Only the British Library owns a manuscript with narrative scenes. As cited above, this was purchased at auction in 1979. This may suggest that such manuscripts were not to be found in the royal libraries or they would certainly have been taken as loot.

A change in clientele may have also been accompanied by the introduction of new craftspeople to the *Kammavācā* workshops. Through the type of materials used in production, the decorative *Kammavācā* manuscript owes much to the ornamented lacquerware of Burma. Yet in these narrative scenes, it is also possible to see the influence of wall painting and *parabaik* illustration. This might demonstrate the redeployment of different craftspeople following the dissolution of the Burmese court. The interest in storytelling is something of a departure from the static depictions of animals and *Devas* of traditional *Kammavācā* ornamentation. The experimentation with perspective indicates an awareness of contemporary art trends. The producer must have been familiar with the standard depictions of the *Mahānipāta* and life of the Buddha stories, along with possessing proficient ability to reproduce those standards on a much smaller scale. The craftspeople, therefore, required not only a mastering of the *shwe-zawa* technique of gold-leaf application used in these manuscripts but also an extensive knowledge of the painting tradition in Burma.

Conclusion

In addition to the social changes taking place across Burma, manuscripts were going through a period of transition. Support materials were becoming noticeably larger; information about donors and even illustrators began to appear on the works. The traditional *magyi-zi* script was gradually being replaced by the standard round script, and printed versions of the *Kammavācā* began to appear.[31] The introduction of narrative scenes was yet another change to a constantly evolving manuscript.

These manuscripts present many challenges. As only a few examples have been recorded, gathering information on the themes included is difficult. Besides the British Library manuscript, no examples of lower quality have been discovered. Tracing the development of storytelling and drawing skills is, therefore, somewhat problematic. Because the manuscripts appear to be relatively new, there may be further and older examples in private collections, which have yet to be accessed and documented.

The manuscripts, while drawing on themes that are deeply embedded in Burmese culture, appear during a period of transition, both socially and within the craft itself. It is intriguing to wonder to what extent this trend may have continued and indeed evolved. The answer may lie in a private collection yet to be discovered.

Notes

1 I have omitted a manuscript from a Thai monastic collection previously discussed in "In Search of a Mon *Kammavācā*," in Patrick McCormick, Matthias Jenny, and Chris Baker, eds., *The Mon over Two Millennia: Monuments, Manuscripts, Movements* (Bangkok: Institute of Asian Studies, Chulalongkorn University, 2011). Although the manuscript contains at least two narrative scenes (probably from the *Vessantara Jātaka*), it is not possible to view the full manuscript, and the format and style of illustration are closer to the Thai than to the Burmese tradition of ornamentation.

2 *Magyi-zi*, or tamarind-seed script, is a thick black lacquer script, similar to the square Burmese script. It is rarely seen on anything other than the *Kammavācā* manuscripts.

3 Naomi Appleton, *Jātaka Stories in Theravada Buddhism: Narrating the Bodhisatta Path* (Farnham, Surrey, U.K.: Ashgate, 2010), 73; Catherine Raymond, "Notes on a Burmese Version of the *Vessantara Jātaka*, as Represented on Three *Shwe Chi Doe* in the NIU Burma Art Collection," *Journal of Burma Studies* 16 (2012): 123–48 (125).

4 Appleton, *Jātaka Stories* (note 3), 72–73. As Appleton points out, this is in fact the Buddha-to-be's "antepenultimate birth," as his penultimate birth takes place in Tusita Heaven. See also pp. 72–74 for a discussion on the placing of the *Vessantara Jātaka* within the *Jatakapali*.

5 See Ward, "In Search of a Mon *Kammavācā*" (note 1), for further information on decorative Mon *Kammavācās*.

6 Alexandra Green, "From Gold Leaf to Buddhist Hagiographies: Contact with

Regions to the East Seen in Late Burmese Murals," *Journal of Burma Studies,* 15 (2011): 305–58 (314). This article discusses Burmese contact with Lan Na and Ayutthaya in the eighteenth century.

7 Green, "From Gold Leaf to Buddhist Hagiographies" (note 6), 319–20.

8 Patricia M. Herbert, *The Life of the Buddha* (London: British Library, 1993), 10.

9 Edward B. Cowell, *The Jataka; or, Stories of the Buddha's Former Births* (Pali Text Society, Cambridge University Press, 1895–1907).

10 Raymond, "Notes on a Burmese Version of the *Vessantara Jataka*" (note 3), 125, uses two Burmese-language sources for her study of the Northern Illinois University *Vessantara Jātaka* textile panels.

11 See Alexandra Green, "Deep Change? Burmese Wall Paintings from the Eleventh to the Nineteenth Centuries," *Journal of Burma Studies* 10 (2005): 1–50 (14), for standardization of *Jātaka* representations in temple paintings of the seventeenth and eighteenth centuries.

12 As FPL KV-141 is an incomplete manuscript and is missing a second folio with full ornamentation, we can safely assume that the second folio of the pair would have also shown a scene from the *Vidhūra Jātaka*.

13 According to Raymond, "Notes on a Burmese Version of the *Vessantara Jātaka*" (note 3), 128, most post–eighteenth-century wall paintings in Burma depict only two horses when illustrating the *Vessantara Jātaka*; the NIU textile, however, depicts four Brahmin.

14 Raymond, "Notes on a Burmese Version of the *Vessantara Jātaka*" (note 3), 131. Interestingly, the vessel that appears in the NIU textile and that which appears in the *Kammavācā* closely resemble each other. Raymond suggests that the NIU vessel is an English teapot.

15 Raymond, "Notes on a Burmese Version of the *Vessantara Jātaka*" (note 3), 134; Gillian Green, "Verging on Modernity: A Late Nineteenth-Century Burmese Painting on Cloth Depicting the *Vessantara Jātaka*," *Journal of Burma Studies* 16 (2012): 79–121 (84). Raymond believes that by the eighteenth century, monk's clothing is preferred when depicting Vessantara in the forest. In Green (p. 84), this late nineteenth- or early twentieth-century cloth painting depicts Vessantara in monk's clothes, with his right shoulder bare.

16 Raymond, "Notes on a Burmese Version of the *Vessantara Jātaka*" (note 3), 139.

17 I am grateful to Alexandra Green for her assistance in identifying this scene.

18 Green, "From Gold Leaf to Buddhist Hagiographies" (note 6), 320; Muriel C. Williamson, "The Iconography of Arched Harps in Burma," in *Music and Tradition: Essays on Asian and Other Musics Presented to Laurence Picken*, ed. Richard

R. Widdess and R. F. Wolpert (Cambridge: Cambridge University Press, 1981), 209–28 (219).

19 In Burmese wall paintings, the Bodhisatta is always seated or reclining as he listens to the music. No examples of the Buddha in a standing position have been found (Alexandra Green, personal communication). Identification of other standing versions of this scene may assist in tracing the origins of this narrative illustration.

20 In fact, the *Vidhūra Jātaka* is illustrated three times, but one of the manuscripts, FPL KV-141, is missing its pair, and therefore the direction of its visual narrative cannot be analyzed.

21 The hanging textile panel, B.C. 90.4.276, Panel 3, discussed in Raymond, "Notes on a Burmese Version of the *Vessantara Jātaka*" (note 3), contains the same *Vessantara* scenes as the second folio of Private Collection ms. 1. Here again the narrative moves from right to left, suggesting that the movement may be intended to reflect the children returning to their original home.

22 See Green, "Deep Change?" (note 11), 5, for the use of dividers in wall paintings.

23 I am grateful to U Thein Lwin and U Maung Maung Thein for their assistance with the translation of captions.

24 *Thayo* is a thick paste made from resin, ash, and clay that is used to mold ornamentation or hold small fragments of colored glass in place on the covers of a manuscript.

25 As a printed version of the *Kammavācā* text is now preferred for monastic use, many old manuscripts are being sold back to the *Kammavācā* makers for reuse. Manuscripts on an ivory support are being cleared of their text, awaiting a new commission in round script.

26 Appleton, *Jātaka Stories* (note 3), 125, references Brown and Dehejia: Vidya Dehejia, "On Modes of Visual Narration in Early Buddhist Art," *Art Bulletin* 72 (1990): 374–92; Robert L. Brown, "Narrative as Icon: The *Jātaka* Stories in Ancient Indian and Southeast Asian Architecture," in Juliane Schober, ed., *Sacred Biography in the Buddhist Traditions of South and Southeast Asia* (Honolulu: University of Hawaii Press, 1997), 64–109.

27 See Appleton, *Jātaka Stories* (note 3), 128, 131, and 137, for a discussion of Brown's and Walters's identification of a "symbolic role" in the *Jātaka* images: Brown, see note 28; Jonathan S. Walters, "Stupa, Story, and Empire: Constructions of the Buddha Biography in Early Post-Asokan India," in *Sacred Biography in the Buddhist Traditions of South and Southeast Asia*, ed. Schober (note 26), 160–92.

28 Green, "From Gold Leaf to Buddhist Hagiographies" (note 6), 325.

29 Green, "Deep Change?" (note 11), 16.

30 Appleton, *Jātaka Stories* (note 3), 146.

31 A printed *Kammavācā*, from Hamsawaddy Press, dated 1892, can be found in the British Library collection, Asia, Pacific & Africa ORB99/28. The format of the printed version currently produced imitates that of a lacquered manuscript.

PART II

Inscribing Religious Practice and Belief

CHAPTER 4

Drawn to an "Extremely Loathsome" Place

The Buddha and the Power of the Northern Thai Landscape

Angela S. Chiu

An old tradition in northern Thailand holds that the Buddha visited the region in his lifetime, deploying his extraordinary powers to fly through the air from India accompanied by a vast entourage of disciples. He visited many villages, delivering sermons for the benefit of local inhabitants and resting and receiving alms along the way at the region's famous mountains, caves, and lakes. In each place, the Buddha left personal items, usually strands of his hair and imprints of his feet on the ground, to serve as objects of veneration for his new devotees. He also predicted the rise of the great Buddhist cities of the region of Lan Na and commanded that after his death, or *parinibbāna*, remains of his body, his relics,[1] be transported from India and enshrined in these great places.

Buddhist legends and historical accounts such as these, called *tamnan* (chronicles), have been written down since at least the fifteenth century in present-day northern Thailand, the heart of the Lan Na cultural region which extended into northeast Burma, southwest China, and northwest Laos. The texts were inscribed by monks on dried palm leaves that were then bundled and sewed together. Most of these palm-leaf manuscripts were composed in Tai vernacular languages, often Yuan or northern Thai, as well as Khoen, Lao, and Lue, interspersed with Pali; a minority were written only in Pali. The scripts include Lan Na Tham, the most common, Fak Kham, and Thai Nithet. The leaves are not illustrated with pictures. It has been estimated that the number of discrete stories or accounts of *tamnan* is in the low to mid two hundreds.[2] Owing to a tradition of respect for copying, numerous editions of the same texts were inscribed by monks through the centuries. While countless manuscripts have been lost over time to fire, insects, and age, the monastic libraries of northern Thailand are still estimated to hold hundreds of thousands of palm-leaf manuscripts, including *tamnan* and other kinds of texts.[3] Only a small percentage of manuscripts has been catalogued. Today some texts of *tamnan* continue to be reprinted and distributed in the form of paper booklets published by monasteries, scholars, and others.

The accounts in *tamnan* of the visits of the Buddha to the region and his predictions about Lan Na are not corroborated by Buddhist canonical texts, which make no mention of such travels. Nonetheless, numerous pieces of the Buddha's bodily remains, or relics, are enshrined around the region. John S. Strong has eloquently highlighted that the Buddha's relics are "*extensions* of the Buddha's biography," which travel to distant places and help legitimize empires and spread the religion (Pali: *sāsana*) to places the Buddha probably never visited himself.[4] Canonical accounts relate that the Buddha's funeral at Kusinārā (the present-day Indian town of Kushinagar) in the fifth century BCE ended with the division and distribution of his remains to eight royal families of ancient India. Two centuries later, the great Buddhist King Asoka is said to have gathered and then redistributed the relics to stupas (reliquary monuments) he constructed in 84,000 locations across Asia. Strong evocatively remarked that Asoka "centripetalized" the relics, "distribut[ing] them outward in a way that allows them to exist locally, while at the same time emphasizing their ongoing ties to the center, their ultimate unity and interconnection."[5] The Lan Na chronicles integrate the region's relics into this grand diffusion of the Buddha's remains.

While Strong's interpretation emphasizes the view from the center of Buddhism in ancient India, the Lan Na *tamnan* may be said to provide the view from the "periphery," in the sense of Thailand's physical distance from India as well as the position of the Lan Na chronicles outside of the Buddhist canon. At the same time, in the chronicles, the relation between "center" and "periphery" is a nuanced one. A striking aspect of the Lan Na relic stories is their persistent emphasis on place. The *tamnan* strongly underscore the association of relics to what is called in Thai their *thi yu, sathan,* or *than*, the place of enshrinement or establishment, the site, often a mountain, where devotees typically bury the relics deep in the ground.

It is through Lan Na's special sites that the Buddha's agency is activated and expressed on an ongoing basis. This essay highlights how the motifs, themes, and structure of *tamnan* reflect this relationship of Buddha and place, which entails not a "conquest" of Lan Na by Buddhism but the process in which Buddhist forms and values were integrated by Lan Na people with their own interests and sense of the landscape. To provide a perspective on Lan Na manuscripts in the sphere of Buddhist literature, this essay will also make some comparison with the conceptions of place in Buddhist canonical texts and in manuscripts of the Lankan and Burmese-Mon traditions that are historically related to that of Lan Na.

The Chronicle of the Buddha Image Lying on a Mango-Tree Log

To provide an example, let us look at the *Tamnan Phra Non Khon Muang* (Chronicle of the Buddha Image Lying on a Mango-Tree Log). This story recounts the origin of a large reclining statue of the Buddha at Wat Phra Non Khon Muang, a monastery located in Mae Rim district in Chiang Mai in northern Thailand. A version of this story is embedded in the nineteenth-century *Chiang Mai Chronicle* and suggests that the statue is connected to King Kawila (reigned 1782–1816).[6] Wat Phra Non Khon Muang itself printed the *tamnan* as a booklet in 1958; this version closes with the date 1816.[7] Thus the composition of the *tamnan* probably dates to the early nineteenth century.

The following summary is based on a palm-leaf manuscript inscribed at Wat Banluk in Mueang Nga subdistrict of the Mueang district of Lamphun city in 1962. It is now in Bangkok in the collection of the Siam Society and has been digitized by the École française d'Extrême-Orient.[8] The text begins during the lifetime of the Buddha Gotama. It relates that for a time, Gotama left India to go on an extended journey around the world spreading Buddhism by distributing his hair relics and giving teachings. One day he arrived at the Ping River, near the present-day site of Chiang Mai, which at that time had not yet been built. The Buddha thought, "The previous Buddha Kassapa in the past came to the base of a great mango tree here. And now I should go to the tree" (p. 1A).[9]

By that time, the mango tree had been there a very long time until it had fallen down and died, and the log was lying on the ground. The Buddha was in pain after eating some bad pork, so he lay down on top of the log. His disciples came over and said, "This place is extremely loathsome. Why does the Lord Buddha come to lie down here?" The Buddha replied that in the past, the previous Buddha Kassapa had also come to this spot. The Buddha then declared to his students that his teachings to them were completed. He predicted that after his death, monks would bring his ashes and enshrine them here and that a monument of the Buddha sleeping on a mango-tree log would be built (pp. 1A–1B).

Just then a *yakṣa*, or ogre, named Alawaka, approached Gotama and demanded to know who he was.[10] Gotama introduced himself as the Buddha, but the ogre was skeptical and said, "How come the Buddha has such a small body?" For in the past, Alawaka had met the Buddha Kassapa, who was big and tall. Gotama then magically transformed himself to the same height as Kassapa. The ogre seeing this raised his hands in reverence. The Buddha told Alawaka that he would give him teachings if the ogre built a pavilion in which the Buddha could sit and also brought him some water for bathing. Alawaka called to his friends, two pigs, to find clean and pure water. The ogre himself went and gathered lotus leaves to make the pavilion. The Buddha gave teachings to the ogre, and (because of the ogre's improvement) one of the ogre's teeth fell out. The Buddha stroked his head and gave a strand of his hair to Alawaka. He told Alawaka to place the hair relic in the same spot where the ogre had previously put a hair relic of the Buddha Kassapa (pp. 1B–2B).

The Buddha Gotama then addressed his disciples. He told them about past events that had happened in that place by the mango tree. Long ago, a king called Phraya Kumphamittarat received a gift of eight mangos. He gave his gardener, Settawaka, the mango pits to plant, and the mango tree grew, with branches turned toward the four directions like an umbrella. The king later handed the gardener some silver and gold to bury for safekeeping, and the gardener interred them by the tree. Gotama noted that one of his current disciples had been Phraya Kumphamittarat in a previous life, while Alawaka the ogre had been Settawaka the gardener in an earlier existence. Alawaka had also met the Buddha Kassapa, who had been traveling around benefiting beings in every settlement, at the mango tree. Kassapa had given teachings to Alawaka and presented the ogre with a hair relic, which the ogre, at Kassapa's order, had buried under the mango tree (pp. 2B–4B).

The Buddha Gotama then prophesied that eight hundred years after his death, or *parinibbāna*, Alawaka would be reborn as a human monarch and raise up the *sāsana* by building a Buddha image and placing it on that site. Two thousand years after the Buddha's *parinibbāna*, the two pigs would be reborn as kings, and the ogre, again reborn as a monarch, would support the religion. The Buddha also called on the god Indra to safeguard Buddhism in that place (p. 5A).

Gotama then left the tree and ascended a mountain, where he gave teachings to some villagers. He predicted the founding of Chiang Mai in the place. The Buddha observed that the gods in heaven were so happy about the prediction that they caused golden rain to fall all over the earth (p. 5B).

The Buddha then noticed an abandoned village nearby. The local people said that the villagers had run away because there were two ogres, a husband and wife, who liked to eat humans. The Buddha spoke to the ogres, telling them that his *sāsana* would be established there. He convinced them to stop eating people and to protect Buddhism in return for the Buddha's blessing (p. 6A).[11]

The Buddha then traveled on to benefit beings in other settlements, eventually going to the base of two *rang* trees in Kusinārā in ancient India, where he entered *parinibbāna* (p. 6B).

In the last part of the chronicle, the predictions made by the Buddha at the mango-tree log are all said to come to pass. Eight hundred years after the Buddha's *parinibbāna*, monks arrived in northern Thailand carrying ashes of the Buddha to enshrine at the mango tree, and an image of the Buddha lying on the fallen tree was also created and established on the spot. Two ogre's teeth containing hair relics and the silver and gold were discovered in the ground and were enshrined there. After that, the country was prosperous and peaceful. The monks predicted that Buddhism would flourish in the city. The chronicle's author ends the text noting the blessings of good health and security that will be received by anyone who sponsors copies of this chronicle (pp. 6B–11A).

The Tamnan Phra Non Khon Muang *and the Connection Between Buddha and Place*

As the title of the manuscript suggests, the text recounts the origin of the famous Buddha image, a shrine for a "relic of commemoration" of the Buddha.[12] However, in terms of its structure and objective, the chronicle might be more aptly described as the history of a particular place. It is basically an account of the attraction of the Buddha to a special location, the future site of the city of Chiang Mai. The structure of the narrative is reminiscent of that of a *jātaka*, or story of a previous existence of the Buddha, in which the narrative begins with the Buddha "in the present" and weaves together threads of past and future. Through this structure, the chronicle generates a multitemporal resonance of agencies centered on place.

The *Tamnan Phra Non Khon Muang* situates the Buddha's journey to Tai territories late in his lifetime, not long before his death in Kusinārā. The events in Chiang Mai in fact echo some of the events of the final days of the Buddha's life as recounted in Buddhist canonical sources. Perhaps this is not surprising, as the statue of the Buddha Image Lying on a Mango-Tree Log is in a reclining posture, which in iconographical terms usually signifies either the Buddha resting or in the midst of his *parinibbāna*. The northern Thai chronicle mentions the Buddha in both contexts, but this particular reclining image has a more complex meaning, which is indicated by the elaborate chronicle, as well as by the relationship between the chronicle and the canonical text that describes the Buddha's last moments in detail, the *Mahāparinibbāna Sutta*.

The chronicle and the *sutta* share several correspondences. They both feature the Buddha resting in the proximity of mango trees, near Chiang Mai in the chronicle, and at the mango grove of Cunda the smith near Pāvā in the *Mahāparinibbāna Sutta*. Both texts depict the Buddha feeling sick after eating some kind of pig-related food, and in both texts, he informs his disciples that he has nothing further to teach them. In the chronicle and in the *sutta*, the Buddha chose to spend time in places that on the surface seemed very unattractive. In the chronicle, the Buddha rested in a spot that his disciples complained was "extremely loathsome," but the Buddha pointed out that the previous Buddha Kassapa had visited there. In the *Mahāparinibbāna Sutta*, the Buddha laid down to die in Kusinārā, which his chief attendant, Ānanda, ridiculed as a "little town of wattle-and-daub, right in the jungle in the back of beyond!" The Buddha explained that Kusinārā was once a great capital called Kusāvatī, ruled by King Mahāsudassana, a wheel-turning monarch.[13] The *Mahāsudassana Sutta* expands this discussion and reveals that the Buddha was King Mahāsudassana in one of his past existences, in fact the last of seven rebirths he had had in that same place.[14] The connection of the Buddha to the "little town of wattle-and-daub" is hence through this illustrious past life. In the *Tamnan Phra Non Khon Muang*, the Buddha is connected to the place by noting that a predecessor in the lineage of Buddhas had visited there. In both the chronicle and the *sutta*, the Buddha's actions are presented as being motivated by past situations that had occurred in a particular place.

In the *tamnan*, it was through the place that Kassapa's agency was manifested. The character of the site of the mango tree as a place of action over many periods is emphasized by the visits of the two Buddhas, the establishment on the spot of multiple relics and treasures, and the cycles of rebirths of humans, ogres, and animals repeatedly circling around the place. Even the establishment of Chiang Mai is presented as a case of suitable positioning, as the founding of the city is connected to the attractions of a place where the presences of Buddhas and their relics were activated; it is striking that this account of the origin of the city does not mention Mangrai, the originator of Lan Na's first royal dynasty who famously founded Chiang Mai in the year 1296.

Place in Other Northern Thai Chronicles

A number of northern Thai chronicles about bodily relics or Buddha images also emphasize the close association between the Buddha and place. For example, the *Chronicle of Wat Suan Dok*,[15] which may have been composed in the early to mid-sixteenth century, recounts the discovery and enshrinement of relics by King Kuena of Chiang Mai (reigned 1355–1385) and Sumana, a monk who seems to have played a leading role in establishing Lanka's Buddhist tradition in Tai territories. The chronicle relates that Sumana, after receiving hints from a local deity, retrieved a Buddha relic from an abandoned stupa in Sachanalai. He brought the relic to show the king of Sukhothai. The relic did not perform any miracles during the royal audience, and so Sumana deduced that the relic should not be enshrined in Sukhothai. He conveyed it to Chiang Mai. There the relic flew into the air, emitting a luminous glow, and miraculously duplicated itself. One relic was enshrined in a stupa built by King Kuena. The other was mounted on the back of a royal elephant. The elephant was set loose, as it was assumed that deities would help guide the elephant to the place where the relic should be enshrined. The elephant climbed up a nearby mountain, reaching the summit; it was so beautiful there that the monk Sumana and King Kuena thought that the summit must be the place for the relic. Nonetheless the elephant did not stop and continued onward, going up another mountain, Doi Suthep (Pali: Ucchupabbata). At the top, the elephant cried out loudly three times, marched around in a circle three times, and then, crouching on its knees, died on the mountain. Thus, it was known that the mountaintop was the spot where the relic should be enshrined. The chronicle highlights how the site of enshrinement was not decided by monks or kings but by the relic itself.

Another *tamnan* that underscores the significance of place for Buddha relics is the *Chronicle of the Great Relic of Hariphunchai* (Pali: Haripuñjaya; modern Lamphun).[16] This story, which probably dates to the fifteenth century, concerns the establishment of what has historically been one of Lan Na's most venerated sites. The chronicle recounts that King Athittarat built a new palace, and while enjoying the royal outhouse there, the king was attacked by a crow, which informed him that a Buddha relic had been buried long ago under the bathroom. The king tried to move the relic to a new location, but the more people dug on the spot, the further down the relic dove into the ground, and Athittarat had to enshrine it on the spot. As in the *Chronicle of Wat Suan Dok*, the relic displayed autonomy; it would only move if it allowed itself to be moved, and it knew where it should be enshrined. Like other Lan Na chronicles, the story indicates that each action or representation of the Buddha is paired with a unique, specific place.

Place in the Life of the Buddha

It may be that the significance of place for the Buddha was a theme that northern Thai devotees found particularly compelling in their interpretation of accounts of the Buddha's life originating from canonical texts, which also inspired sermons, ritual recitations and mural paintings. In the *Mahāparinibbāna Sutta*, the Buddha famously named his places of birth, Enlightenment, first sermon, and final *nibbāna* as sites of pilgrimage,[17] suggesting a powerful relationship between the Buddha's actions and place. Such a relationship is also evoked in the *Nidānakathā* of the *Jātakaṭṭhakathā*, traditionally said to have been written by Buddhaghosa in Lanka in the fifth century and now found as the introduction to the collection of *jātakas*. The *Nidānakathā* recounts events in the life of the Bodhisatta (Buddha-to-be) Siddhattha–Buddha Gotama up until the donation of the Jetavana monastery. Place served an essential function in the development of Siddhattha's Buddhahood. For instance, Siddhattha, while seated under a banyan tree, was given an offering of milk rice in a golden bowl from Sujātā, a landowner's daughter who mistook him for a tree deity. Siddhattha finished the meal and is said to have carried the golden bowl toward the bank of the river Nerañjara. He waded into the river through a ford that had also been trod by previous Buddhas-to-be on the days of their Enlightenment. When the Bodhisatta set the bowl on the surface of the water, it glided against the current and then plunged down to the river bottom to settle precisely beneath the submerged bowls of the three previous Buddhas.[18] This and other important biographical events in Siddhattha's life preceding his Enlightenment are depicted as identical to those that happened in the lives of previous *bodhisattas*. Siddhattha or Gotama is fitted into the lineage of his predecessors through repetition of actions that are specified as occurring in specific places.

After leaving the river, Siddhattha situated himself beneath the Bodhi tree, where he proceeded to attain Enlightenment amidst ferocious attacks by the *devaputta* Māra and his army. Māra demanded, "Rise, Siddhattha, from that seat. It is not meant for you. It goes to me." The Buddha-to-be replied that Māra had not practiced the Perfections or fulfilled other missions and that therefore "this seat is not meant for you. I alone have the right to it."[19] The accumulation of achievements, built up through many lifetimes of a Buddha-to-be, is metaphorically represented by the seat.

The seat is described as a specific physical place. The *Nidānakathā* recounts the Bodhisatta's approach to the space under the Bodhi tree. The text explains that before settling down, he stood at three different sides of the tree in turn and considered that each was probably not the place of Enlightenment. Finally, he sat down in a cross-legged seated position at the eastern side. This was the spot that "trembles not and shakes not," where all previous Buddhas had meditated. The Bodhisatta Siddhattha "realized that that was the stable place, never forsaken by any of the Buddhas, the seat for the destruction of the aggregate of defilements."[20]

The reliable stability of the spot turned out to be necessary to support Siddhattha amid the storms of swords, hot ash, mud, and other substances that Māra hurled at the meditating sage. Indeed, after his success, the *Nidānakathā* reports that the Buddha said, "This, my throne of victory, is a unique throne. Whilst seated here all my aspirations have come to pass; and I will not rise from it yet." The Buddha even remained there for such a long time after the Enlightenment that some deities suspected that he had "not yet given up his attachment to the throne of victory"; to prove them wrong, the Buddha performed a double miracle.[21]

Other texts also highlight the significance of the seat to the Buddha's Enlightenment. In reviewing descriptions of the Enlightenments of Gotama and of the series of previous Buddhas as recounted in various canonical Pali texts, I. B. Horner observed:

> This *bodhi-pallaṅka* position is uniform for all Buddhas and it is obligatory for them to sit in the same place, *ekasmiṃ yeva ṭhāne*. It is the only one able to support the weight of a Buddha's attainment [Jā. IV. 229], the *jayapallaṅka* [cross-legged position for victory, Jā. I. 77] of all Buddhas, the navel of the earth [Jā. IV. 232f., Mhbv. 79]. No one, not even Sakka himself, can pass over the *Bodhi-maṇḍa* (ground, circuit, platform or dais round a Bodhi-tree) whether a Tree be growing there at the time or not, for here it is that all Buddhas have routed all the defilements.[22]

Siddhattha and the other *bodhisattas* are depicted as being impelled to seek Enlightenment at a certain spot, with the understanding that the Enlightenment could only be realizable at that spot and not at any other. Enlightenment is not only an experience of person but also of place. A singular unification of person and place is essential to the arising of this rare experience. In turn, the attainments of Enlightenment by the Buddhas in their seats made manifest the specialness of these places. As Horner observed, even Sakka, the king of the gods, became compelled to honor these sites. Hence what occurred in these places continued to exert agency even after the Buddhas stood up from the seats and walked away. These places were channels for Buddhahood.

Lankan Manuscripts About Relics

The structure, motifs, and themes of Lan Na *tamnan* also have correspondences with those of Lankan texts about relics, such as the chronicle of Lanka's famous Tooth Relic (*Dāṭhāvaṃsa*), traditionally said to have been written in the fourth century in a Sinhalese language and translated into Pali in the thirteenth century by Dhammakitti Thera.[23] The *Jinakālamālīpakaraṇaṁ*, a Lan Na chronicle thought to date to the early sixteenth century, provides a recounting of the spread of Buddhism from India to Lanka and then to mainland Southeast Asia and includes summaries of the history of the tooth and of other Lankan relics.[24] Copies of the

Tooth Relic chronicle and of other Lankan texts, as well as newly written texts claiming a Lankan heritage, may still be found today in Lan Na's monastic libraries. This dissemination of Lankan texts may have originated in the fourteenth and fifteenth centuries, when Tai monks journeyed to Lanka to study at the island's monasteries, where they were reordained in Lankan orders that they then propagated in Tai territories after they returned home. Lankan literature at that time was a flourishing tradition, having enjoyed a renaissance that began in the mid-twelfth century and continued for at least two hundred years.[25]

Both Lankan and Lan Na stories bridge temporal and geographical separations in order to make the Buddha "present" after his death or *parinibbāna*. They recount how the Buddha made visits to particular places outside ancient India where he made prophecies about his relics that were later fulfilled through the efforts of monks and lay devotees. The texts also highlight the activities of past Buddhas to emphasize the long and profound historical significance of particular sites. The visits of three previous Buddhas to Lanka and their depositing relics there are recounted in the earliest known Lankan chronicle, the fourth- or fifth-century *Dīpavaṃsa*[26] as well as in the fifth- or sixth-century *Mahāvaṃsa*.[27]

This trope of visits of previous Buddhas to various places, according to Étienne Lamotte, should be understood as part of a propagandizing effort by Buddhist missionaries.[28] Strong likewise noted that advancing the claim that the sacred sites of non-Buddhist cultures are associated with past Buddhas "would seem to be have been an ideal way for incorporating non-Buddhist, pre-Buddhist, or Brahmanical elements into the Buddhist fold."[29] Indeed, Dhida Saraya has posited that in Lan Na, beginning in the fourteenth century, kings deliberately enshrined Buddha relics at the sites of local guardian deities around Lan Na in order to establish the rulers' rights and powers over the lands and devotees of those deities.[30] The significant role of divinities in many relic stories may evoke the submission of local deities to the Buddha. On the other hand, the subsuming of local guardian sites to Buddhism may not have been an objective of all Lan Na relic establishments; as we have seen in the stories, the site that turns out to be a relic's residence may be described as unremarkable, disgusting, or difficult to find, suggesting that it was not a recent place of veneration.

The multitemporality found in Lan Na *tamnan* echoes that found in Lankan texts. The *tamnan* depict Buddhas, people, and gods as connected across time through karma; in the *Tamnan Phra Non Khon Muang*, for instance, Gotama encountered the same ogre who had met previous Buddhas; the converted ogre later enjoyed a rebirth as a king. Writing of Lankan chronicles, Jonathan S. Walters suggested that these sorts of karmic connections entail interweaving two dimensions of time. One dimension of time is "calculable" (Pali: *saṅkheyya*), referring to the dating of historical events such as the birth and other life events of specific kings. A second dimension of time is "incalculable" (*asaṅkheyya*), referring to time-places that are beyond mapping, such as kings who lived before the first king and places that existed before the creation of the world. Through the workings of karma, circumstances of "calculable" time could be connected to practices of infinite, cosmic tradition: specific Lankan kings and monastics were portrayed in the *Dīpavaṃsa* and *Mahāvaṃsa* as heirs to earlier great Buddhist empires because of their meritorious actions over incalculable previous Buddhist eras. These kings and monks were thus specially empowered to enable the people of Lanka to progress in their individual journeys toward Buddhahood.[31] To Walters's analysis, it should be added that the operation of karma across multiple dimensions of time occurs through fixed place. It is in place—in Anurādhapura on Lanka, in Chiang Mai, in Hariphunchai—that the accumulation and repetition of actions are effected through time and given power in the present.[32]

Burmese-Mon Manuscripts About Relics

Ways of understanding and discussing Buddha relics and images also circulated between Lan Na and Burmese areas, particularly the Mon region of lower Burma, through connections of trade and religious exchange over a number of centuries.[33] The exchange of Buddhist manuscripts seems likely to have increased after 1558, when kings from Burmese territories conquered Lan Na. They exercised suzerainty over the region for more than two hundred years.

In fact, the most common *tamnan* discovered in a 2007 survey of selected

Lan Na monastic libraries by researchers of the École française d'Extrême-Orient is the *Tamnan Phra Kesa That Singkuttara Thakong*, the story of the Buddha's hair relics enshrined in contemporary Yangon at the Shwedagon, today Burma's most revered Buddhist monument.[34] The chronicle's account in northern Thai language (Yuan) seems to closely resemble the story of the hair relics as presented in Mon chronicles found in Burma.[35] The Burmese-Mon story is itself apparently an extension of a Pali tradition about two merchants, Trapuṣa and Bhallika, who donated food to the Buddha and became his first disciples. In the Burmese-Mon account, the Buddha granted hair relics to the pair and commanded that they be enshrined on Mount Singuttara, where relics of the three previous Buddhas were installed. Trapuṣa and Bhallika faced various difficulties on the way to finding the place, whose location was unknown to them. Finally, after years of searching, they received help from the gods and reached the mountain, where they enshrined the relics amid the manifestation of many miracles, such as the blossoming of a hundred thousand lotus flowers, the swaying of the mountains, and the curing of the blind and the hunchbacked.

As in the Lan Na chronicles described previously, the agencies of past Buddhas are invoked, highlighting the lineage of Gotama and the long history of Mount Singuttara's importance. The performance of miracles signaled the greatness of the unification of relic and place and demonstrated the power of the Buddha vested in the relics. The difficulties encountered by the two merchants in their quest and the various kinds of assistance from deities take up the main part of the text and enhance the sense of the relics' arrival at Singuttara as an extraordinary occasion. In a trope familiar in the Lan Na chronicles, Singuttara is highlighted as being visually rather ordinary, and it is only after gods clear the top of it of vegetation that the two merchants take notice of it. Perhaps the affinities of this relic story with those of Lan Na contributed to its popularity in Lan Na. The reliquary site is not mentioned in the early sixteenth-century *Jinakālamālīpakaraṇaṁ*, although Trapuṣa and Bhallika are, so perhaps the story spread in Lan Na after the Burmese takeover in the mid-sixteenth century.

Numerous Burmese stories of relics of the Buddha may be found in *The Glass Palace Chronicle of the Kings of Burma* (1923), Pe Maung Tin's translation of part of the *Hmannan Mahayazawindawgyi*. Written in 1829, the *Hmannan* is a history of Burma prepared at the command of King Bagyidaw (reigned 1819–1837) by a committee of monks, ministers, and Brahmins based on earlier manuscripts and inscriptions.

Here let us consider one example of a relic story that provides an interesting comparison with Lan Na stories. *The Glass Palace Chronicle* recounts that King Aniruddha (reigned ca. 1044–1077) invaded the city of Śrī Kṣetra. While there, he broke a hole into a shrine and retrieved the Buddha's frontlet (forehead bone; Pali: *uṇhīsa*). He conveyed it back to his capital of Bagan. There he mounted the relic on the back of an elephant that he allowed to wander freely. The huge animal eventually knelt on a sandbank. Upon seeing the chosen site of the relic, the king "was sorry, for he had thought the religion would last full five thousand years. His heart was ill content that the white elephant knelt not upon natural soil, but on shifting sand." But Sakra (Sakka), king of the gods, entered Aniruddha's dreams that night and assured him that the *sāsana* would indeed endure for five millennia. "And Sakra strengthened the ground with solid rock, two hundred and forty thousand times thicker than before, and clamped it all around with iron plates." Aniruddha then had the Shwezigon pagoda built on the spot to house the relic, which performed miracles before its enshrinement.[36]

The unstable sandbank was seen to bode badly for the life span of the *sāsana*, which according to tradition was believed to be five thousand years. The story thus indicates the broader purpose of enshrining Buddha relics: to support the security of Buddhism. As in some of the Lan Na stories, the relic in this case ended up in a place that seemed to be quite inappropriate for its enshrinement. Of course, it was because the relic had to be installed in its chosen home that Sakka's engineering feat became necessary. The deity's assistance also reflects a universal hierarchy in which the Buddha is venerated by all beings, including gods.

Aniruddha is still today revered in Burma as one of the country's greatest kings, and his behavior in enshrining the relic is depicted in the text as exemplary. *Hmannan Mahayazawindawgyi* was intended by King Bagyidaw to provide a "standard" for kings and the conduct of affairs of state and religion.[37] Bagyidaw ordered its compilation five years after his kingdom's humiliating defeat by the British Indian Army in the First Anglo-Burmese War. The chronicle directly linked Bagyidaw's "somewhat upstart" Konbaung dynasty to the prestigious lineages of the kings of Bagan and of the Sakyas, the family of the Buddha Gotama.[38] The chronicle's accounts of the kings' ability to handle relics helped connect the power, righteousness, and significance of the Burmese court to a wider and deeper Buddhist history.

This story of the frontlet relic also reflects Burmese conceptions of place and the nature of divine landscape. A Burmese belief held that a Buddha relic or image, or the palace of a king, needed to be situated in a place "strong" enough to support such a powerful entity. The location must be one demonstrated to be a "strong" place or "victory ground" (Burmese: *aungmyei* or *zeyabbhumi*; Pali: *jayabhūmi*), a place of triumph, in past and future, over various kinds of forces. This concept also influenced the siting of the Burmese capital city, which was not fixed in place but could be transferred to a new site deemed more auspicious by the ruling monarch. For example, King Alaungpaya (reigned 1752–1760) referred to the royal capital city of Shwebo as a "victory ground where all enemies are overcome" and noted with satisfaction that there his small band of soldiers had prevailed over a much larger Mon army. The king's advisers then drew a connection between Shwebo and another victory ground, Kapilavatthu, the capital of the Buddha Gotama's Sakyan clan. Thus the royal city of Shwebo was understood to be heir to the lineage of important sites of ancient Buddhist India.[39] The local tradition of royal courts asserting that each capital city was the subject of prophecy by the Buddha Gotama may have begun in the late eleventh or twelfth century or earlier.[40] During the Konbaung period (1752–1885), court officials listed the names of cities that had existed on the site of the royal capital during the periods of past Buddhas.[41] Royal courts also performed rites to affirm the identification of the city as the latest and greatest capital. This included the ritual collection of portions of earth from previous royal cities and other symbolically significant places and the deposition of the earth at the new royal city. During the establishment of Amarapura in the late eighteenth century, earth was collected from four each of lakes, islands, and hills as well as from sixteen former royal palaces in cities across the Konbaung empire. The earth was then ritually deposited into the ground at the site of the future royal palace. This ceremony, as François Tainturier suggested, could have been intended to associate the geographically significant places of Burmese history with the sixteen *Mahājanapadā*, or polities, of ancient Buddhist India.[42]

Thus the concept of "victory ground" joined together king, capital, and *sāsana*. For the Burmese, "strong" could be a symbolic as well as a literal characteristic: the very particles of the ground were understood to have a certain power. This explains the reactions of Aniruddha and Sakka to the sandbank.

There may be a connection to Lankan ideas about earth. The *Mahāvaṃsa* and *Thūpavaṃsa* list in detail the layers of materials spread to form the foundation of the Great Stupa, which is now called the Ruwanwelisaya, in Anuradhapura. These include crushed stone, brick, and iron, as well as *marumba*, a type of earth transported by monks from the Himalayas, and "butter clay," described as the clay in the place where the divine river Gaṅga dropped down to the earth, "where varieties of rice of spontaneous growth arise."[43] As in the Burmese examples, earth from significant places was brought to support an important monument.

In Lan Na, the term "victory ground," *jayabhūmi* (Thai: *chaiyaphum*), also appears in *tamnan*. For example, the *Chiang Mai Chronicle*, a history of Chiang Mai composed around 1828 and based on earlier texts, describes King Mangrai's identification in 1292 of a site as a *jayabhūmi* and suitable for the construction of his capital, Chiang Mai. The site was seen as a "victory ground" because a pair of hog-deer, a mother and fawn, astonishingly defeated a pack of wolves on the spot.[44] There does not seem to be any reference to earth-gathering rituals in this account of Chiang Mai's founding. Nonetheless, the significance attached by the Burmese to earth may have been familiar to Lan Na people; in the earth-gathering ritual noted above for Amarapura, among the sixteen sites of former Burmese royal palaces from which earth was collected was Chiang Mai, then considered a Burmese domain.[45]

Conclusion

As Strong has insightfully observed, relics are emanations from the Buddha across time and space that can remedy the problem of the physical absence of the Buddha, express the spread of his teachings, and mark the territories of "conquest" of the *sāsana*. In a similar vein, Donald Swearer characterized Lan Na as a "*buddhadesa*" or land of the Buddha:[46] in stories about the Buddha's historical visits to Lan Na, "the Buddha can be understood as sacralizing the entire region by impregnating it with his bodily relics following an itinerary that is at once an etiological justification for Buddhist pilgrimage sites and a symbolic guarantee of the order, prosperity, and productivity of the land."[47] Thus Swearer highlights Buddhist and universal motifs in *tamnan* to situate Lan Na within a wider sphere of Buddhist history and thought in Asia.

These interpretations, which focus on connecting local places to a canonical center or to a broader world of Buddhist values, highlight how the presence and actions of Buddha relics cause great transformation to the places where they are established. They emphasize the effect of the Buddha's agency on local sites. As we have seen, a more nuanced view of the power of relics is visible through attention to the mode of narrative and description in extant manuscripts of Lan Na chronicles. In *tamnan*, the Buddha and his relics are attracted to Lan Na places. The relics are depicted as confirming or adding to the significance of each place. It is the unique sites of Lan Na that enable the activation of agencies over time. One might then describe the production of *buddhadesa* as the reanimation of Buddha agency in time via place. The *tamnan* speak of the process by which Buddha agency has arisen in Lan Na. This production has an active, eventful sense, in which the interaction between Buddha and Lan Na is not just one-way. Mircea Eliade famously elaborated the concept of "hierophany," which is "the act of manifestation of the sacred."[48] The Lan Na relic stories highlight that the act of manifestation of Buddha agency is enabled by distinctive places for such agency. The narratives, shifting among past, present, and future, depict how the intersection of dimensions of time, the "calculable" history of certain lands and peoples and the "incalculable" history of the Buddhas, is enabled by specific sites.

This essay's comparison of Lan Na, Buddhist canonical, Lankan, and Burmese-Mon traditions, though limited in scope, suggests an interplay of common regional themes and particular local interests. Peter Skilling has observed that in the early and premodern eras, texts arose within "cultural conversation" that was "multi-directional and multi-dimensional" between Tais and other people across Asia. While there are Tai texts that are not known in the literature of other Theravādin centers, there was a complex Asian "intertextuality."[49] In a similar vein Justin McDaniel has characterized Lan Na's monastic manuscript production and consumption as forming "interpretive communities and reading cultures."[50] The development of Lan Na's Buddhist literature did not entail merely a copying of earlier Buddhist examples but involved the selective deployment of structures, motifs, and themes to convey particular understandings, such as the conception of place that suited local objectives and ways of understanding the landscape.

Notes

This essay includes references to manuscripts in the collection of digitized, transcribed, and translated Lan Na manuscripts of the École française d'Extrême-Orient (EFEO) at the Princess Maha Chakri Sirindhorn Anthropology Centre in Bangkok. The digitized manuscripts were made publicly accessible in 2014 at www.efeo.fr/lanna_manuscripts/ (last accessed December 15, 2014). The author is immensely grateful to François Lagirarde, leader of the Lan Na manuscripts project, for generous access to and assistance with the project's resources in 2010. My citations from the manuscripts are based on the project's translations of the manuscripts to central Thai.

1 The canonical *Kāliṅgabodhijātaka* purports that the Buddha identified three kinds of remainders or shrines (*cetiya*) as sites for his veneration: 1) a shrine for a relic of the body (*sarīrikacetiya*), 2) a shrine for a relic of use (*paribhogacetiya*), and 3) a shrine of commemoration or indication (*uddesikacetiya*). Relics "of use" are usually thought to refer to such items as the Buddha's bowl and robe, while relics

"of commemoration" are typically interpreted to mean Buddha images. John S. Strong, *Relics of the Buddha* (Princeton: Princeton University Press, 2004), 19–20.

2 Udom Rungrueangsi has suggested that there are 206 different texts of *tamnan*. François Lagirarde has estimated that there are 230. Udom Rungrueangsi, "Tamnan," in *Saranukrom Watthanatham Thai Phak Nuea* [Encyclopaedia of Culture of the Northern Thai Region] (Krung Thep: Mulaniti Saranukrom Watthanatham Thai, 2542 [1999]), 5:2412–16. François Lagirarde, "Temps et lieux d'histoires bouddhiques: À propos de quelques 'chroniques' inédites du Lan Na," *Bulletin de l'École française d'Extrême-Orient* 94 (2007): 59–94 (60).

3 Sarassawadee Ongsakul, *History of Lan Na*, trans. Chitraporn Tanratanakul (Chiang Mai: Silkworm, 2005), 1–8.

4 Strong, *Relics of the Buddha* (note 1), 7.

5 Ibid., 143.

6 The story in the *Chiang Mai Chronicle* ends with the note that eight brothers will be born and that the city of Chiang Mai will be founded in 1796; these seem to be references to the Chao Chet Ton, the collective name for King Kawila and his brothers, who led the expulsion of the Burmese from Lan Na and resettled the city of Chiang Mai, which had been abandoned since 1776 after a siege by the Burmese. Further, the story appears in the text just after the death of Kawila in 1816. See David Wyatt and Aroonrut Wichienkeeo, trans. and ed., *The Chiang Mai Chronicle*, 2nd ed. (Chiang Mai: Silkworm, 1998), 187–92.

7 *Tamnan Wat Phra Non Khon Muang* (Chronicle of the Monastery of the Buddha Image Lying on a Mango-Tree Log) (Wat Phra Non Khon Muang, 2501 BE [1958]), 12. This publication is a central Thai translation of the *tamnan*.

8 *Tamnan Phra Non Khon Muang* (Chronicle of the Buddha Image Lying on a Mango-Tree Log), Siam Society, Bangkok, EFEO 001 002. Inscription date 2505 BE (1962 CE). Yuan language and Lan Na Tham script.

9 The page numbering used here is not that of the original manuscripts but that assigned by EFEO, which designates verso and recto sides of a folio as A and B.

10 The inspiration for this encounter was apparently the *Āḷavaka Sutta* (*Saṃyutta Nikāya* 10.12 and *Sutta Nipāta* 1.10), in which the ogre Āḷavaka tested the Buddha with challenging questions and became his follower after hearing his responses.

11 This account is apparently related to an old northern Thai legend about Pu Sae and Ya Sae, a pair of human-eating ogres who lived long ago on Doi Suthep, a mountain overlooking Chiang Mai, and were reformed by the Buddha. Pu Sae and Ya Sae are still honored as guardian deities in a special festival in Chiang Mai today.

12 At Wat Phra Non Khon Muang today, there is no evidence of the log nor of a reproduction of it. The log itself could be described as a "relic of use" (*paribhogacetiya*).

13 Maurice Walshe, *The Long Discourses of the Buddha: A Translation of the Dīgha Nikāya* (Boston: Wisdom, 1995), 266.

14 Ibid., 289–90.

15 *Tamnan Wat Suan Dok* (Chronicle of the Flower Garden Monastery), Wat Pa Sak Noi, Sankhamphaeng, Chiang Mai, EFEO 002 003; inscription date 1309 CS (1947 CE); Yuan language with some Pali; Lan Na Tham script. Another version also from Wat Pa Sak Noi is *Tamnan Wat Suan Dokmai* (Chronicle of the Flower Garden Monastery), EFEO 002 004; inscription date 1296 CS (1934 CE); Yuan language with some Pali; Lan Na Tham script. See also the English translation of the *Tamnan Phra That Suthep* (Chronicle of the Relic of Suthep), in Donald K. Swearer, Sommai Premchit, and Phaithoon Dokbuakaew, *Sacred Mountains of Northern Thailand and Their Legends* (Chiang Mai: Silkworm, 2004), 69–83. The career of Sumana and the founding of Wat Suan Dok (Pali: Pupphārāma) are also described in the *Jinakālamālīpakaraṇaṁ*: see Jayawickrama, *The Sheaf of Garlands of the Epochs of the Conqueror, Being a Translation of the Jinakālamālīpakaraṇaṁ of Ratanapañña Thera of Thailand* (London: Pali Text Society, 1978), 117–20, 126–27.

16 According to Sarassawadee Ongsakul, the enshrining of the Great Relic of Hariphunchai took place around 1157. Sarassawadee, *History of Lan Na* (note 3), 38. The story of the Great Relic can be found as part of a number of other manuscripts and may be read today in modern publications including translations. The account here is based on *Tamnan Phra That Chao Lamphun lae Tamnan Wat Ton Kaeo* (Chronicle of the Sacred Relic of Lamphun and Chronicle of Wat Ton Kaeo), Siam Society, Bangkok, EFEO 001 005, 21B–25B; inscription date 1275 CS (1913 CE); Yuan language with some Pali; Lan Na Tham script. The story is also included in the *Jinakālamālīpakaraṇaṁ* and *Mūlasāsanā* chronicles of Lan Na; see, for example, Jayawickrama, *The Sheaf of Garlands* (note 15), 106–9; and Sommai Premchit and Donald K. Swearer, "A Translation of *Tamnān Mūlasāsanā Wat Pā Daeng*: The Chronicle of the Founding of Buddhism of the Wat Pā Daeng Tradition," *Journal of the Siam Society* 65, no. 2 (1977): 73–110 (77).

17 Walshe, *Long Discourses of the Buddha* (note 13), 263–64.

18 N. A. Jayawickrama, trans., *The Story of Gotama Buddha (Jātaka-nidāna)* (Oxford: Pali Text Society, 1990), 92–93.

19 Ibid., 97.

20 Ibid., 94.

21 Ibid., 103.

22 I. B. Horner, trans., *The Minor Anthologies of the Pali Canon,* Part 3: *Chronicles of Buddhas (Buddhavaṁsa) and Basket of Conduct (Cariyāpiṭaka)* (London: Pali Text Society, 1975), xxxix.

23 Bimala Churn Law, *On the Chronicles of Ceylon* (Calcutta: Royal Asiatic Society of Bengal, 1947), 22–23.

24 These include the southern branch of the Bodhi tree, the right eyetooth relic, and the forehead (frontal) bone. See Jayawickrama, *Sheaf of Garlands* (note 15), 65–81, 88–94.

25 Kate Crosby, "The Origin of Pāli as a Language Name in Medieval Theravāda Literature," *Journal of Buddhist Studies* 2 (2004), 70–116 (95–96).

26 Strong, *Relics of the Buddha* (note 1), 43.

27 Wilhelm Geiger, *The Mahāvaṃsa, or the Great Chronicle of Ceylon* (Colombo: Ceylon Government Information Department, 1960), 99–109.

28 Étienne Lamotte, *History of Indian Buddhism*, trans. Sara Webb-Boin (Louvain-la-Neuve: Institut Orientaliste, 1988), 338, quoted in Strong, *Relics of the Buddha* (note 1), 41.

29 Strong, *Relics of the Buddha* (note 1), 41.

30 Dhida Saraya, *The Development of the Northern Thai States from the Twelfth to the Fifteenth Centuries* (Ph.D. dissertation, University of Sydney, 1982), 115–17.

31 Jonathan S. Walters, "Buddhist History: The Sri Lankan Pāli Vaṃsas and Their Commentary," in Ronald Inden, Jonathan Walters, and Daud Ali, *Querying the Medieval*, 99–164 (102ff) (New York: Oxford University Press, 2000).

32 One significant theme in the Lankan chronicles did not seem to find similar favor in Lan Na. Kevin Trainor has pointed out that Lankan narratives are structured to reflect a "hierarchy of possession" in which control over relics is directly correlated to individual attainment of the Buddhist ideal of nonattachment. Thus monks are privileged as those who hold power over access to relics. In Lankan relic stories, monks are depicted as those who shepherd the relics to Lanka, while kings have the duty to enshrine them. Kevin Trainor, *Relics, Ritual, and Representation in Buddhism: Rematerializing the Sri Lanka Theravada Tradition* (Cambridge: Cambridge University Press, 1997), 134. In the Lan Na chronicles, those who access and manage relics include not only monks but also laypersons and gods. Moreover, monks (such as Sumana in the *Tamnan Wat Suan Dok*) do not necessarily display magical skills, as they so often do in the Lankan chronicles, in the handling of relics.

33 The *Cāmadevivaṃsa*, a fifteenth-century chronicle about Hariphunchai, states that 250 years after the founding of the city (i.e., perhaps in the eleventh century), a cholera epidemic broke out and the survivors fled temporarily to lower Burma, where the inhabitants spoke the same language. Donald K. Swearer and Sommai Premchit, *The Legend of Queen Cāma: Bodhiraṃsi's Cāmadevīvaṃsa: A Translation and Commentary* (Albany: State University of New York, 1998), 105–6.

34 François Lagirarde, "Temps et lieux d'histoires bouddhiques" (note 2), 67.

35 For a history, analysis, and summary of several texts found in Burma, see Pe Maung Tin, "The Shwe Dagon Pagoda," *Journal of the Burma Research Society* 24, no. 1 (1934): 1–91.

36 Pe Maung Tin and G. H. Luce, trans., *The Glass Palace Chronicle of the Kings of Burma* (London: Oxford University Press, 1923), 87–88.

37 Pe Maung Tin, "Introduction," in Pe Maung Tin and Luce, *Glass Palace Chronicle* (note 36), ix–xxiii (ix).

38 Thant Myint-U, *The Making of Modern Burma* (Cambridge: Cambridge University Press, 2001), 81.

39 Tun Aung Chain, "Prophecy and Planets: Forms of Legitimation of the Royal City of Myanmar," in Tun Aung Chain, *Selected Writings of Tun Aung Chain*, 124–50 (124–28). (Yangon: Myanmar Historical Commission Golden Jubilee Publication Committee, 2004). The area of Shwebo is still in contemporary times famous for its association with the victories of the Konbaung dynasty, and several well-known Buddha relics and monuments are located there. Pilgrims who visit Shwebo often collect small amounts of earth in containers that they then place on their household altars for protection from threats of all kinds. Juliane Schober, "Mapping the Sacred in Theravada Buddhist Southeast Asia," in *Sacred Places and Modern Landscapes*, ed. Ronald A. Lukens-Bull, 1–27 (16) (Tempe: Monograph Series Press, Program for Southeast Asian Studies, Arizona State University, 2003).

40 François Tainturier, *The Foundation of Mandalay by King Mindon* (Ph.D. dissertation, University of London, 2010), 45–46. See also Tun Aung Chain, "Prophecy and Planets" (note 39), 126.

41 Tainturier, *Foundation of Mandalay by King Mindon* (note 40), 63–64.

42 Ibid., 39–42.

43 Geiger, *Mahāvaṃsa* (note 27), 191–92; and N. A. Jayawickrama, trans., *Chronicle of the Thūpa and the Thūpavaṃsa* (London: Luzac, 1971), 100–1.

44 Wyatt and Aroonrut, *Chiang Mai Chronicle* (note 6), 41–47.

45 Tainturier, *Foundation of Mandalay by King Mindon* (note 40), 41.

46 Donald K. Swearer, "Signs of the Buddha in Northern Thai Chronicles," in *Embodying the Dharma: Buddhist Relic Veneration in Asia,* ed. David Germano and Kevin Trainor, 145–62 (157) (Albany: State University of New York Press, 2004).

47 Donald K. Swearer, "Part I. Interpretation," in Swearer and Sommai, *The Legend of Queen Cāma* (note 33), 3–34 (10).

48 Mircea Eliade, *The Sacred and the Profane*, trans. William R. Trask (Orlando: Harcourt, 1987), 11.

49 Peter Skilling, "Geographies of Intertextuality: Buddhist Literature in Pre-modern Siam," *Aséanie* 19 (2007): 91–112 (98–101).

50 Justin McDaniel, "Two Buddhist Librarians: The Proximate Mechanisms of Northern Thai Buddhist History," in *Buddhist Manuscript Cultures*, ed. Stephen C. Berkwitz, Juliane Schober, and Claudia Brown, 124–39 (137) (London: Routledge, 2009).

CHAPTER 5

Shifting Modes of Religiosity

Remapping Early Chinese Religion in Light of Recently Excavated Manuscripts

Ori Tavor

Recent years have witnessed a surge in publications concerning newly discovered manuscripts from the Warring States period (453–211 BCE). The discovery of such excavated materials, which offer us fresh and unmediated access to previously unseen sources, has prompted a call for a reassessment of previously held convictions regarding the intellectual world of this constitutive era in Chinese history. In this essay, I will draw on two texts from the Shanghai Museum manuscript corpus in an attempt to reassess the world of religious discourse in early China and reconstruct the evolution of two competing modes of religiosity accompanied by distinct theories of ritual efficacy.

The Shanghai Museum corpus includes more than 1,200 inscribed bamboo strips purchased in 1994 by the Shanghai Museum on the Hong Kong antiquities market.[1] So far, nine transcribed volumes have been published by the Shanghai Museum Press between 2001 and 2012. Written in archaic script that was identified as originating in Chu [楚], an ancient southern state located in the area of modern-day Hunan and Hubei provinces, these manuscripts, which include alternative versions of texts that were preserved in the received tradition, such as the *Classic of Changes* [*Yijing*, 易經, also known as the *Zhouyi*, 周易] and many previously unknown texts, were assigned the approximate date of 300 BCE. The strips, which are uniform in width but range in length from 24 to 27 centimeters, were originally bound together through pared notches. Given that the cords themselves did not survive, and since unlike earlier Warring States manuscripts from the Chu region, such as the Guodian [郭店] corpus, the Shanghai Museum texts were not archaeologically excavated, organizing the strips into coherent textual units proved to be a long and highly problematic process.[2]

Following an initial surge of enthusiasm among the sinological community, a growing number of scholars have begun raising some questions regarding the value of studying such unprovenanced manuscripts.[3] In this article, I will argue that despite the difficulty in reading these sources as coherent textual units, when read against the backdrop of the received sources, the ideas articulated in them can help us obtain a more nuanced picture of early Chinese intellectual discourse. In this particular case, when read alongside the writings of the Confucian philosopher Xunzi, the Shanghai manuscripts reveal a lively debate between two modes of religiosity: a practical theology associated with a mechanical approach to ritual and a new elite mode of religiosity that traced the power of ritual to a moral theology accompanied by a fixed body of ritual practices. Studying these excavated materials is thus crucial in understanding the intricate changes that took place in the religious world of pre-imperial China and also reveals new avenues of continuity regarding the role of early Chinese religious discourse in the formation of organized religion.

Early Chinese Religiosity and the Spring and Autumn Ritual Reorientation

Recent archaeological excavations suggest that one of the main forms of ritual activity in early Chinese religion involved interaction with ancestral spirits and other divine powers through the mediums of sacrifice and divination. Excavations in sites associated with the Bronze Age cultures of Erlitou [二里頭, first half of the second millennium BCE] and Erligang [二里岡, mid-second millennium BCE] point to the existence of standardized ritual practices concerned with the proper disposal of the dead. In the Anyang site [安陽], the location of the capital of the late Shang Dynasty [商, ca. 1200 BCE], archaeologists have unearthed large caches of turtle plastrons and bovine scapulae that were used in divination rituals. The inscriptions carved on these oracle bones reveal the existence of a complex ritual system accompanied by a specialized vocabulary and strict schedules.[4]

While oracle bone inscriptions contain much information about the rituals of Shang religion, they lack any overt theoretical discussions on the nature, origin, and function of these practices. This, however, does not mean that Shang religion was a religion devoid of any theological framework. It merely indicates that the religion lacked an explicit theology. In fact, the divinatory and sacrificial practices described in the oracle bones correspond to what Jan Assmann refers to as an embedded implicit theology. As opposed to an explicit theology, which operates on a reflective distance from religious practice, explaining it on a theoretical level, in an implicit theology, the ritual acts themselves gave meaning to action by categorizing, constellating, and differentiating among various aspects of reality.[5]

Shang religion thus poses a challenge to scholars who wish to unravel the implicit theology embedded in its rituals and present it to the modern reader. In order to facilitate this process, they often resort to modern theories of ritual and sacrifice. Two such influential interpretations of Shang religious practices, David Keightley's depiction of Shang religion as the "making of ancestors" and Michael Puett's notion of the "give-and-take" [*do ut des*] mentality, are heavily influenced by the commerce model of sacrifice first introduced by French sociologist Marcel Mauss. According to Mauss, sacrifice follows the same rules as gift exchange and thus can be seen as a binding contract between man, the gift giver, and deity, the receiver. By offering a victim in the form of a sacrifice, the sacrificer purchases the powers of the deity for a given price.[6] While the details of Mauss's model have been criticized, the commerce metaphor and the notion of ritual as a negotiation technique that facilitates communication and exchange between the human and supernatural realms are still used by scholars of religion.[7]

Keightley draws on this model in arguing that the complex ritual system of the Shang reflects a conscious attempt to deal with the capricious nature of the supernatural world through the construction of a standardized bureaucratic religious hierarchy that follows fixed patterns of ritual interaction. This mode of religiosity, he argues, sees the relationship between the human and the divine as negotiable. By addressing the deceased by their proper name and entering them into the sacrificial schedule, the worshiper is able to take an unpredictable and potentially dangerous ghost and make it into a proper ancestor.[8]

Drawing on Keightley's scheme, Puett describes the religious system of the Shang as one of continuous negotiation in which ritual interactions were used to "influence, mollify, and determine the will of the divine powers, to persuade them to grant assistance and to prevent them from making disasters." Furthermore, this give-and-take mentality was still a fundamental component of Western Zhou Dynasty [周, 1045–771 BCE] religiosity. Western Zhou religious hymns and bronze inscriptions, for instance, were designed to build a proper ancestral pantheon that would work on behalf of the Zhou royal house, while the rituals that accompanied them present an attempt to convince the ancestors to descend to the human world and provide the performer with divine blessings.[9]

During the Spring and Autumn period [770–481 BCE], while the nominal sovereignty of the Zhou kings was still generally accepted, the de facto control over the territories of the Zhou state moved to the hands of local rulers. These sociopolitical changes were accompanied by a shift in the religious sphere, which has been described as a "reorientation away from the ancestors."[10] Although the spirits of the ancestors are still mentioned in Spring and Autumn ritual bronze inscriptions, they are no longer depicted as the addressees of sacrifice or potential givers of aid. The focus of ritual action thus shifts from the veneration of ancestral spirits to the self-panegyrical glorification of the living members of the community, and ritual efficacy is depicted as a result of the descendant's own ritually correct behavior.[11]

The notion of Spring and Autumn "ritual reorientation" has been described as an important initial step in the emergence of a new theory of ritual that found full articulation in the Warring States period. This era is often referred to as the age of the "Hundred Schools of Thought," a time in which new ideas about the self and its relationships with sociopolitical institutions found articulation in a growing corpus of philosophical literature. In the early twentieth century, Western-educated Chinese scholars, influenced by evolutionary models advocated by Victorian anthropologists James Frazer and Edward Tyler,[12] began characterizing the Warring States period as an important junction in which magical and religious modes of thinking gave way to rational philosophical systems such as Confucianism and Mohism.[13] Despite the eventual demise of the evolutionary model in the second half of the twentieth century, its effects on contemporary interpretations of intellectual discourse during the Warring States period are still palpable.

Yuri Pines, for example, argues that the reorientation in ritual practice in the Spring and Autumn period was accompanied by an intellectual reappraisal of the old Zhou ritual system and the traditional sociopolitical order it represented. Drawing on a variety of received textual sources, Pines describes a process of distillation of certain normative aspects of ritual from a loose set of religious sacrifices to a fully developed idea of a ritual system [*li*, 禮], a set of ethical and sociopolitical guidelines that functioned as the source of political legitimacy and as the means of perpetuating internal social cohesiveness.[14]

Pines provides the following account from the *Zuo Commentary of the Spring and Autumn Annals* [*Chunqiu Zuozhuan*, 春秋左传, henceforth referred to as the *Zuozhuan*] in order to substantiate the emerging distinction between the philosophical notion of ritual as a secular set of ethical and social principles and the court rituals associated with the old religion of the Zhou. In this account, when Duke Zhao of Lu [魯昭公, r. 541–510 BCE] is visited by Duke Ping of Jin [晉平公, r. 557–532 BCE], the latter was thoroughly impressed by Zhao's performance of the proper court ceremonies. When he professed his admiration to his advisor Nü Shuqi [女叔齊], Nü replied that Duke Zhao's behavior did not reflect his proficiency in ritual [*li*] but only his knowledge of ceremony [*yi*, 儀], claiming that:

> Ritual is that by which [a ruler] protects his State, carries out his governmental decrees and does not lose his people. Now the command over the government [of Lu] is at the hands of the clans, and he [Duke Zhao] cannot take it [from them]. . . . His [royal] house is divided into four parts, and his people get their food from others, not thinking of him or taking any consideration for his future. He

> is a ruler whom calamity visits personally, and yet he has no regard to what is proper for him to do. The root and branches of ritual lie in these sorts of things, yet [the duke] fusses over trivial ceremonies as if they were of utter importance. Is it not far from the truth to say that he is good in ritual?[15]

According to Pines, this passage clearly indicates that the meaning of *li* as a pattern of governance has overshadowed its religious origins. Moreover, drawing on received sources such as the *Analects* [*Lunyu*, 論語], the *Mencius* [*Mengzi*, 孟子], and the *Mozi* [墨子], Pines argues that the transformation of the semantic field of *li* culminated in the work of the late Warring States thinker Xunzi [荀子, ca. 310–218 BCE], who was able to distill its "pure essence" and finally disassociate *li* from its ties to the Zhou ritual system.[16]

Pines's characterization of this process as a departure from the Zhou religious framework derives from the nature of the primary texts he uses, which are all received sources.[17] Archaeological excavations conducted in the second half of the twentieth century, however, have unearthed a significant number of texts that were never transmitted into the received literary corpus. For scholars of early Chinese religion, the discovery of manuscripts in sites such as Mawangdui [馬王堆], Zhangjiashan [張家山], and Shuihudi [睡虎地] revealed the existence of a flourishing literary tradition composed of technical manuals that offer us a glimpse into the realm of popular religiosity. These texts cover a wide range of topics ranging from self-cultivation manuals to divination and exorcism handbooks. Analysis of these manuscripts suggests that the authors of these texts, namely, astrologers, physicians, diviners, and other ritual specialists who are often grouped together under the rubric of "natural experts," were active participants in the Warring States intellectual scene; their literature functioned as an important vehicle for the transmission of ideas from popular religion to elite philosophical discourse.[18]

The discovery of newly excavated manuscripts has driven leading sinologists to call for a reassessment of early Chinese intellectual history.[19] In the following pages, I will argue that a close reading of two excavated texts from the Shanghai Museum corpus, *Drought of the Great King of Jian* [*Jian*

Dawang Bohan, 柬大王泊旱] and *Great Drought of Lu* [*Lubang Dahan*, 魯邦大旱], reveals the existence of a tension between two types of theology accompanied by distinct theories of ritual efficacy when compared with the backdrop of the received literature. These texts suggest that the Warring States reconceptualization of ritual signaled the emergence of a new moral religiosity that eventually became a fundamental component of Chinese organized religion.

Competing Modes of Religiosity: Two Examples from the Shanghai Manuscripts

The emergence of philosophical disputation in China is often associated with the changing sociopolitical reality of the Warring States period and the rise of a new influential social group known as scholar-officials [*shi*, 士]. This new group viewed civil service as their route to power and influence; and, in most cases, they served as advisors and even high-level functionaries in the governments of the feuding Warring States.[20] The emergence of new players in the political arena resulted in an inevitable power struggle between the new aspiring elite and the old guard. Anecdotes of disputes between the two sides, especially criticism directed toward the efficacy of the rain sacrifices associated with the old Zhou religious framework, are abundant in the received literature, such as the following passage from the *Zuozhuan*: "There was a great drought in [the state of] Zheng. [The king] sent three of his officials to perform a sacrifice on Mulberry Mountain. They cut down the trees [for the sacrifice], but it did not rain. Zichan said: "[the goal of] performing a sacrifice on the mountain is to nourish its forests. These [men] have cut down the trees and thus their crime is immense." He proceeded to take away their official positions and fiefdoms."[21] Zichan [子產, also known as Gongsun Qiao公孫僑, d. 522 BCE] is mentioned throughout the *Zuozhuan*, alongside such figures as Nü Shuqi and Yan Ying [晏嬰, d. 500 BCE], as critics of popular religious ideas.[22] Despite the relative abundance of such anecdotes, it is important to note that the king's instinctive reaction to the drought was to send his ritual specialists to perform a sacrifice on top of the sacred Mulberry Mountain.[23] This suggests that despite Zichan's criticism, the technical give-and-take mentality identified by Keightley and Puett as the dominant theory of ritual efficacy in Shang and Western Zhou religion was still the most natural reaction to a state of crisis in the Warring States period.

The excavated Shanghai manuscript *The Great Drought of Lu* contains a similar narrative of this tension between different modes of religiosity. When a great drought occurred in the state of Lu, Duke Ai [魯哀公, r. 494–468 BCE] summoned Confucius and pleaded for his advice. Confucius, in return, explained that the drought was caused by the duke's failure to practice moral government. When asked for a concrete solution to the problem, Confucius provided the following statement: "The common people only know of the *shuo* rainmaking sacrifice[24] [directed towards] the spirits but know nothing of cultivating moral government. Thus, you must be generous in offering jades and silks to the [Spirits of the] Mountains and Rivers and also practice moral government."[25] Confucius's recommendation to pursue both courses resonates with his famous assertion in the *Analects* regarding the need to venerate ghosts and spirits but to keep them at a distance.[26] The text, however, does not end with that. Upon his return, Confucius reports the case to his disciple Zigong [子貢, 520–446 BCE] and asks for his opinion. Zigong's response is quite surprising:

> Practicing moral government, thereby serving Heaven above, this is correct! Lavishly offering jades and silks for the [Spirits of the] Mountains and Rivers, this I cannot endorse. As for mountains, stones are their skin and trees are their people. If the sky does not send down rain, the stones will roast and the trees will die. Their desire for rain is certainly deeper than ours—how can they rely solely on our words [of evocation]? As for rivers, water is their skin and fish are their people. If the sky does not send down rain, the water will dry up and the fish will die. Their desire for water is certainly greater than ours—how can they rely solely on our words [of evocation]?[27]

Modern scholars, such as Liu Lexian [劉樂賢], attempt to resolve this harsh criticism directed toward Confucius by one of his own disciples by raising the hypothesis that Confucius only suggests the sacrifices to the spirits as a public gesture to pacify the common people who are incapable of understanding the real cause behind the drought: the ruler's failure to practice moral government.[28] While Liu's attitude might be dismissed as contemporary Confucian apologetics, his analysis does raise two important points. First, the tension between the natural experts, who sought to perform the rainmaking sacrifices in order to appease the mountain and river spirits, and their opponents, who believed that serving Heaven can only be achieved through moral government, was important enough to be recorded and preserved in this text. Second, this passage confirms that the give-and-take theory of ritual belonged to the realm of popular religion. The newly educated elite, represented in this case by Zigong, sought to criticize this mode of religiosity and replace this implicit practical theology with a new, explicit moral theology based on a devotion to a standardized set of ethicoreligious guidelines rooted in cosmic principles.[29]

So far, based on these passages alone, Zichan's and Zigong's criticism can be understood as representing the process described by Pines as an attempt to distill a normative secular sociopolitical system from a religious framework that had lost its relevance. However, when read against the backdrop of *The Drought of the Great King of Jian*, another excavated text from the Shanghai Museum corpus, an alternative explanation arises. According to this text, when a severe drought fell upon his kingdom, the ruler of Jian, a territory inside the larger southern state of Chu [楚],[30] ordered one of his diviners to figure out which deity was responsible for the drought so that they might offer a sacrifice to it in the proper place and stop the drought. The king insisted on participating in the divination process while standing in the blazing sun, and this caused him to fall ill.[31] Taking his illness as another indicator for the dissatisfaction of the deities, the king was greatly distraught and attempted to persuade his diviners to look for an alternative site for the sacrifice. His idea of performing sacrifices to the mountain and river spirits that resided outside the kingdom of Jian, however, attracted much criticism in the royal court. In order to solve this dispute, the rival sides sought the advice of the chief minister. After hearing both side of the argument, he responded:

> Please go back and convey these words to the king. Tell him that from today he will start to recover from his illness. . . . The king is a good ruler. He did not change the fixed rules of divination for his own sake. You, diviner, control the [sacrifices] to the ghosts and spirits in the state of Chu. You also did not dare to change the fixed rules only for the sake of your ruler thereby creating disorder among the ghosts and spirits. Shang Di, the ghosts, and the spirits are highly discerning. They will surely recognize this. Thus, from this day, the king will start to recover from his illness.[32]

Similarly to *The Great Drought of Lu*, this passage suggests that the most natural reaction to a state of drought at the time was to perform a rainmaking sacrifice directed at natural deities. In addition, it also informs us of the structure of these rituals, the identity of the ritual specialists who performed them, and the religious model that underlies them. According to this practical mode of religiosity, the sacrificial procedure begins with a divination designed to ascertain the identity of the responsible deity and locate the appropriate location for the sacrifice. Ritual is thus perceived as a repertoire of techniques placed at the disposal of the ritual specialist in order to create a sacred space in which interaction with the divine is possible. The ultimate success of the sacrifice depends on the ritualist and his ability to use his repertoire to manipulate the deities into reciprocating. This type of trial-and-error style of practical theology associated with Shang and Western Zhou religiosity was still quite pervasive during the Warring States period.[33]

As *The Drought of the Great King of Jian* suggests, however, the practical model advocated by natural experts and ritual specialists was criticized by the new, aspiring elite of *shi*, who offered their own model of ritual efficacy focused on piety to a fixed ethical system of practice. Much like Nü Shuqi, Zichan, and Zigong, the chief minister stresses the overall devotion to the system as a whole over the performance of a specific ritual. His reasoning, however, makes it hard to read his argument as a philosophical distillation indicating a process of secularization. The state of Chu, he argues, has fixed rules about sacrifice. Changing them for the sake of the king's selfish wish for divine blessings will not only harm him politically but will also create chaos in the divine realm. Devotion to this holy fixed system of rituals, however, will not escape the eyes of the High God Shang Di and other supernatural powers.[34] These deities will repay such religious piety by healing the king and, by extension, his state. This argument thus reveals the emergence of a moral theology that links the efficacy of sacrifice to a sustained adherence to a strict system of rituals. The fullest and most mature articulation of this model can be found in the writings of Xunzi.

Patterns of the Way: Xunzi and the Question of Ritual Efficacy

Throughout much of Chinese history, the writings of the Confucian thinker Xunzi were rejected by the cultural mainstream.[35] In the twentieth century, however, a renewed interest in his writings emerged. The same Western-educated Chinese scholars who associated the Warring States period with a shift from religion to philosophy hailed Xunzi as a staunch critic of religion and the forebear of rationalist thought in China.[36] One of the best examples for this attitude is the "Discourse on Heaven" [Tianlun, 天論] chapter, which is said to be the fullest systematic version of the philosophical skepticism and critical attitude toward popular religion exhibited by *Zuozhuan* thinkers such as Zichan and Nü Shuqi.[37] Consider, for example, the following passage: "When stars fall and trees cry, all the people in the state are afraid. They ask: why is this happening? I answer: for no particular reason. Those things occasionally occur due to the transformation of Heaven and Earth and the transformation of *yin* and *yang*. We may be surprised by them, but we should not fear them. Solar and lunar eclipses, unseasonable rains and winds, and dubious sightings of strange stars—these things have been quite common throughout the ages."[38] This passage and the "Discourse on Heaven" chapter as a whole are often read as an attack on the practical mode of popular religiosity. Robert Eno, for example, argues that Xunzi's critique is directed toward the magical mentality exhibited by natural experts such as shamans and other diviners-sorcerers, specifically their claims for transcendental knowledge and ritual authority.[39] Edward Machle, however, depicts Xunzi's attempt to differentiate between the ritual system of *li* and superstitious rituals as religiously motivated. In an attempt to discredit Xunzi's image as antireligious, Machle argues that while the latter represent a result-oriented magical mentality, the *li* are religious since they entail a lifelong commitment to a particular way of life and a detailed theology.[40]

Despite his efforts to contest Xunzi's image as a rationalist and an antitheist, Machle's use of the evolutionary model is problematic. Reading Xunzi against the backdrop of the excavated Shanghai drought texts, however, offers us an opportunity to contextualize his theory of ritual within the Warring States religious discourse. According to this reading, Xunzi's attitude represents a mature articulation of an emerging moral theology accompanied by a fixed body of religious practices known as the system of *li*. Set against the practical theology of the natural experts, this elite mode of systematic religiosity seeks to create an indissoluble link between ritual as a system of ethical and sociopolitical guidelines and its divine cosmic origin. Xunzi's critique of the popular theory of ritual efficacy is presented in the following passage from the "Discourse on Heaven" chapter:

> If a rainmaking sacrifice is held, and then it rains, what of it? I say, there is no reason. It would still rain even if we do not hold the sacrifice. When the sun and moon are eclipsed, a sun-saving rite is performed; when Heaven sends a drought, a rainmaking sacrifice is performed; before deciding upon serious matters, tortoise shell and milfoil divinations are performed. These [rituals] are not held in order to get a result, but in order to establish a pattern. Thus, the gentleman takes [ritual] as a matter of establishing a pattern while the common people take it as a matter of [sacrificing to the] spirits. To take [ritual] as creating a pattern is auspicious. To take it as [sacrifice to the] spirits is ill-fated.[41]

It is important to note that Xunzi does not object to the performance of these rituals but to the religious mentality that underlies them. As opposed to the popular give-and-take mode of religiosity in which rituals are performed for the sake of the spirits, Xunzi's model targets an elite audience in the form of the Confucian gentleman [*junzi*, 君子] and portrays ritual participation as an activity that establishes a pattern. Writing for a new elite audience of educated scholar aspirants, Xunzi wishes to establish a new mode of religiosity based on an absolute sense of devotion to the system of *li* and the ethicoreligious values it represents. Being a gentleman, he argues, involves an enduring commitment to a fixed regimen of ritualized physical, emotional, and spiritual cultivation.[42] Ritual is thus portrayed as one of the most important human activities: "In Heaven, there is nothing brighter than the sun and moon. On Earth, there is nothing brighter than water and fire. Among material objects, there is nothing brighter than pearls and jades. In the human realm there is nothing brighter than ritual and propriety. . . . Therefore, the fate of man lies in Heaven and the fate of the state lies in ritual."[43] As we recall, Xunzi's theory of ritual is depicted by Pines as the epitome of the distillation of *li* from its old religious framework. Nevertheless, Pines also draws our attention to the cosmological and ontological dimension of ritual in Xunzi's thought.[44] Throughout the "Discourse on Heaven" chapter, Xunzi stresses that calamities do not arise because of malicious supernatural powers and thus cannot be averted through ritual activity performed under the mind-set of a practical give-and-take mode of religiosity. The only method for avoiding these calamities, he argues, is to understand the patterns and movements of reality and then to use this acquired knowledge to one's advantage. This connection between human behavior and cosmic patterns, as observed in the notion of moral government [*xingde*] found in *The Great Drought of Lu*, thus reaches full articulation in Xunzi's theory of ritual and the Way [*dao*, 道]: "Those who cross waterways mark them where it is deep. If the markers are not clear, then people will drown. Those who govern people mark the Way. If the markers are not clear, then disorder will arise. Ritual is the marker. Opposing ritual means throwing the world into darkness. Casting darkness upon the world will bring great disorder."[45] Rituals, argues Xunzi, are not arbitrary. They are markers left by sages that function as a prescriptive script, a guiding light for the rest of humanity to follow. Moreover, since rituals are based on the fixed patterns of the Way, one must adhere to them without attempting to alter them. Xunzi's attitude concerning the Way can thus be best understood as one of religious reverence or devotion. By creating an indissoluble link between the structure of the universe and the system of *li*, Xunzi offers an explicit theological justification for a new mode of elite religiosity focused on a commitment to a body of ethicoreligious behavioral guidelines. According to this moral theology, rituals are not performed in order to seek an anticipated result from a supernatural deity. Instead, the performance of rituals of the Way is a pattern-establishing activity that denotes the religious devotion and the moral stature of the practitioner.

Conclusion: Moral and Practical Theologies in Chinese Religions

A close reading of newly excavated manuscripts from the Shanghai Museum corpus against the backdrop of received sources suggests that the Warring States ritual reorientation signified the emergence of a new mode of elite religiosity that challenged the practical implicit theology of the natural experts. The rise of moral religiosity, however, did not signal the disappearance of the give-and-take mentality. While the architects of the new imperial religion of the Han Dynasty [206 BCE–220 CE] used the moral theological framework to construct an official state cult in which the emperor's performance of grand sacrifices is seen as instrumental to the maintenance of social, political, and cosmic harmony, the amoral practical theology associated with the Warring States ritual experts did not die out. Recent studies of Han religion clearly demonstrate that personal religious practices aimed at obtaining individual practical benefit continued to flourish on all levels of society, from the common people to the emperor. Emperor Wu of the Han [漢武帝, 156–87 BCE], for instance, employed the services of natural experts to redesign several state rituals to promote his personal quest for immortality, while commoners often used their services as healers, diviners, and exorcists.[46]

After the fall of the Han, the tension between practical and moral theologies manifested itself in the rhetoric used by the followers of a new religious movement, Celestial Masters Daoism [Tianshi Dao, 天師道], who attempted to undermine the popularity of their religious rivals: local cults. One of the main strategies used by the Celestial Masters in asserting the superior efficacy of their rituals was to claim that their system of practice was based on a moral theology sent down to Earth by the deified Way. By establishing a link between the bureaucratic organization of the celestial realm and their earthly ethicoreligious codes, the Celestial Masters were able to identify their system as orthodox [*zheng*, 正] and the rituals of rival local cults as heterodox [*xie*, 邪].[47]

The tension between these two modes of religiosity that emerged in early China became one of the key features of Chinese religious discourse and was instrumental in the subsequent formation of such organized religious traditions as Daoism and Buddhism. Unlike these institutional religions, however, early Chinese religion is a particularly amorphous entity that does not conform to contemporary definitions of religion, since it lacks many of the features modern scholars view as fundamental, such as a canonical set of sacred scriptures, organized clergy, or a fixed pantheon. In fact, the label "early Chinese religion" does not refer to a specific empirical singularity. It is mainly used as a heuristic device, a term coined by later scholars to help make sense of a collection of phenomena. This, however, does not mean it should be set aside in favor of the study of religious traditions such as Buddhism and Daoism. Recent years have, in fact, witnessed a surge in book-length monographs devoted to early Chinese state cults,[48] funerary practices and visions of the netherworld,[49] self-cultivation and individual pursuits of immorality,[50] as well as an imposing two-part edited volume dealing exclusively with religious beliefs and practices between the Shang and Han Dynasties.[51] Recently excavated manuscripts feature heavily in these studies, as such sources offer us a glimpse into the realm of popular religiosity in early China and reveal a richer picture than the one reflected in the received literary tradition. Studying these manuscripts thus offers us new ways of understanding the religious traditions that are still practiced in the Chinese cultural sphere today and, on a broader level, can contribute to our current definition and understanding of religion as a universal phenomenon.

Notes

1 Ma Chengyuan 馬承源, *Shanghai Bowuguan Cang Zhanguo Chuzhushu*上海博物館藏戰國楚竹書 (Shanghai: Shanghai Guji Chubanshe, 2001–2012), 2001: 1–4.

2 For more information about the excavation of the Guodian corpus, see Wang Chuanfu 王傳富and Tang Xuefeng 湯學鋒, "Jingmen Guodian Yihao Chumu 荊門郭店一號楚墓," *Wenwu* 文物 (1997.7): 35–48.

3 Paul R. Goldin defines an unprovenanced text as "one whose original location is unknown." See Goldin, "*Hengxian* and the Problem of Studying Looted Artifacts," *Dao: Journal of Comparative Philosophy* 12 (2013): 153–60.

4 Robert Thorp, *China in the Early Bronze Age: Shang Civilization* (Philadelphia: University of Pennsylvania Press, 2006), 102–4, 172–85.

5 Jan Assmann, *The Search for God in Ancient Egypt* (Ithaca, NY: Cornell University Press, 2001), 12.

6 Marcel Mauss, *The Gift: The Form and Reason for Exchange in Archaic Societies* (New York: Routledge, 1990), 15–17.

7 Claude Lévi-Strauss, for example, criticizes the directness of the gift-giving scheme and argues for an alternative theory that stresses the role of the sacrificial object as a mediator between the sacred and profane realms. According to this model, the interaction between these two distinct realms is made possible by the annihilation of the sacrificial victim, which creates a vacuum that must be filled by the anticipated benefit (Lévi-Strauss, *The Naked Man* [New York: Harper & Row, 1966], 224–26). Walter Burkert's model of "silent trade," on the other hand, emphasizes the uncertainty of this interaction, arguing that ritual space is constructed in the hope of coaxing the otherwise illusive deity to show itself (Burkert, *Creation of the Sacred: Traces of Biology in Early Religions* [Cambridge, MA: Harvard University Press, 1996], 139–55). Maria Heim questions the validity of applying Mauss's model in studying the practice of *dāna*, almsgiving, and argues that ideal gift relationships in South Asian religious discourse do not take the form of a give-and-take mentality but are instead based on "one-way regard and respect" (Heim, *Theories of the Gift in South Asia: Hindu, Buddhist, and Jain Reflections on Dana* [New York: Routledge, 2004], 54).

8 David Keightley, "The Making of the Ancestors: Late Shang Religion and Its Legacy," in *Religion and Chinese Society*, ed. John Lagerwey, 3–63 (Hong Kong: Chinese University of Hong Kong Press, 2004).

9 Michael Puett, *To Become a God: Cosmology, Sacrifice, and Self-Divinization in Early China* (Cambridge, MA: Harvard University Press, 2002), 41, 67.

10 Lothar von Falkenhausen, *Chinese Society in the Age of Confucius (1000–250 BC): The Archaeological Evidence* (Los Angeles: Cotsen Institute of Archaeology, University of California, 2006), 295–97.

11 Gilbert L. Mattos, "Eastern Zhou Bronze Inscriptions," in *New Sources of Early Chinese History: An Introduction to the Reading of Inscriptions and Manuscripts*, ed. Edward L. Shaughnessy, 85–123 (85–88) (Berkeley: Society for the Study of Early China and the Institute of East Asian Studies, University of California, Berkeley, 1997).

12 A comprehensive survey of evolutionary theories of magic, religion, and science can be found in Stanley Tambiah, *Magic, Science, Religion, and the Scope of Rationality* (New York: Cambridge University Press, 1990).

13 Feng Youlan 馮友蘭, *A Short History of Chinese Philosophy* (bilingual edition) 中國哲學簡史 (英漢對照) (Tianjin: Tianjin Shehui Kexueyuan Chubanshe, 2007), 2–4. This attitude was also shared by early Western sinologists. In an article written in the 1940s, Derk Bodde claims that "it is ethics (especially Confucian ethics), and not religion (at least, not religion of a formal organized type), that has provided the spiritual basis of Chinese civilization" (Bodde, "Dominant Ideas in the Formation of Chinese Culture," *Journal of the American Oriental Society* 62, no. 4 [1942]: 293).

14 Yuri Pines, "Disputers of the Li: Breakthroughs in the Concept of Ritual in Pre-Imperial China." *Asia Major* 13, no. 1 (2000): 1–41 (6–7).

15 Yang Bojun 楊伯峻, *Chunqiu Zuozhuan Zhu* 春秋左传注 (Beijing: Zhonghua Shuju, 1990), 1266. Quoted in Pines, "Disputers of the Li," 15.

16 Pines, "Disputers of the Li" (note 14), 15, 40.

17 By the term "received sources," I am referring to texts that have been transmitted through history by means of scribal copying as opposed to excavated manuscripts.

18 Marc Kalinowski, "Technical Traditions in Ancient China and *Shushu* Culture in Chinese Religion," in *Religion and Chinese Society* (note 8), 223–48 (239–40); Donald Harper, "Warring States Natural Philosophy and Occult Thought," in *The Cambridge History of Ancient China*, ed. Michael Loewe and Edward L. Shaughnessy, 813–84 (814) (New York: Cambridge University Press, 1998).

19 Li Xueqin李學勤, *Chongxie Xueshushi*重寫學術史 (Shijiazhuang: Hebei Jiaoyu Chubanshe, 2002); Edward L. Shaughnessy, *Rewriting Early Chinese Texts* (Albany: State University of New York Press, 2006); Michael Loewe and Michael Nylan, eds., *China's Early Empires: A Re-appraisal* (New York: Cambridge University Press, 2010).

20 Cho-yün Hsü, *Ancient China in Transition (*Stanford, CA: Stanford University Press, 1965), 89–92; Yuri Pines, *Envisioning Eternal Empire: Chinese Political Thought of the Warring States Era* (Honolulu: University of Hawaii Press, 2009), 115–19.

21 Yang, *Chunqiu Zuozhuan Zhu* (note 15), 1382.

22 Paul R. Goldin, *Rituals of the Way: The Philosophy of Xunzi* (Chicago: Open Court, 1999), 39–45; Pines, "Disputers of the Li" (note 14), 14–17.

23 These rainmaking ceremonies [*dayu*, 大雩], usually performed in the summer, are mentioned more than twenty times in the *Zuozhuan*.

24 Reading 敓 as 說. This interpretation is suggested by both Liu Lexian and Li Xueqin (Liu Lexian 劉樂賢, "Shangbo Jian Lubang Dahan Jianlun 上博簡魯邦大旱簡論," *Wenwu* 文物 [2003]: 60–61; and Li Xueqin李學勤, Chongxie Xueshushi重寫學術史 [Shijiazhuang: Hebei Jiaoyu Chubanshe, 2002]: 98). In the *Rites of Zhou* [*Zhouli*, 周禮], jade insignia are mentioned as common offerings in sacrifices to the heavenly bodies (Sun Yirang孫詒讓, *Zhouli Zhengyi* 周禮正義 [Beijing: Zhonghua Shuju, 1987]: 1591).

25 Ma, *Shanghai Bowuguan Cang Zhanguo Chuzhushu* (note 1) (2002), 205–6.

26 Yang Bojun 楊伯峻, *Lunyu Yizhu* 論語譯注 (Beijing: Zhonghua Shuju, 2007), 61–62.

27 Ma, *Shanghai Bowuguan Cang Zhanguo Chuzhushu* (note 1) (2002), 207–9.

28 Liu Lexian 劉樂賢, "Shangbo Jian Lubang Dahan Jianlun 上博簡魯邦大旱簡論." *Wenwu* 文物 (2003.5): 60–64 (62–63).

29 The term I translated as "moral government" is *xingde*, 刑德. In terms of its individual components, the word *xing* refers to laws, while *de* is usually translated as virtue or moral power. John S. Major argues that when used together, the phrase *xingde* refers to a technique of moral government that entails the implementation of punishments and rewards according to cosmic cycles of recession and accretion (Major, "The Meaning of Hsing-te," in *Chinese Ideas About Nature and Society: Studies in Honor of Derk Bodde*, ed. Charles Le Blanc and Susan Blader, 286–87 [Hong Kong: Hong Kong University Press, 1987]).

30 Although the ruler is described in the text as "the Great King of Jian," he was actually a feudal lord in charge of a fiefdom inside the larger state of Chu.

31 Ji Xusheng 季旭昇, Yuan Guohua 袁國華, and Chen Siting 陳思婷, *Shanghai Bowuguan Cang Zhanguo Chuzhushu (si) Duben* 上海博物館藏戰國楚竹書(四)讀本 (Taipei: Wanjuan Loutushu Youxian Gongsi, 2007), 75. For more on self-exposure and rainmaking rituals in ancient China, see Alvin Cohen, "Coercing the Rain Deities in Ancient China," *History of Religions* 17, no. 3/4 (1978): 244–65; and Edward Schafer, "Ritual Exposure in Ancient China," *Harvard Journal of Asiatic Studies* 14, no. 1/2 (1951): 130–84.

32 I have followed the bamboo slip arrangement suggested in this volume and the annotations provided by the editors. This arrangement is also supported by Chen Wei, in Chen Wei陳偉, "*Jian Dawang Bohan* Xinyan 《簡大王泊旱》新研," *Jianbo* 簡帛2 (2007): 259–68.

33 Roel Sterckx, "Searching for Spirit: Shen and Sacrifice in Warring States and Han Philosophy and Ritual," *Extrême-Orient, Extrême-Occident* 29 (2007), 23–54 (32–37).

34 The term *Shangdi* [上帝] first appears in Shang oracle bone inscriptions. By the Warring States period, it seems to indicate the highest deity in the religious pantheon and is often synonymous with Heaven [*tian*, 天]. See Robert Eno, *The Confucian Creation of Heaven: Philosophy and the Defense of Ritual Mastery* (Albany: State University of New York Press, 1990), 1; and Herrlee Creel, *The Origins of Statecraft in China* (Chicago: University of Chicago Press, 1970), 493–502.

35 Paul R. Goldin, "Xunzi and Early Han Philosophy," *Harvard Journal of Asiatic Studies* 67, no. 1 (2007): 135–66 (136–37).

36 Feng, *Short History of Chinese Philosophy* (note 13), 232. This is particularly noticeable in post-1949 Chinese scholarship produced in Mainland China, in which Xunzi is often portrayed as a materialist and his method associated with that of modern science. For a detailed survey, see Goldin, *Rituals of the Way* (note 22), 109 n. 6.

37 Goldin, *Rituals of the Way* (note 22), 47.

38 Wang Xianqian 王先謙, *Xunzi Jijie* 荀子集解 (Beijing: Zhonghua Shuju, 1988), 313.

39 Eno, *Confucian Creation of Heaven* (note 34), 142–43.

40 Edward Machle, "Hsün-Tzu as a Religious Philosopher," *Philosophy East and West* 26, no. 4 (1976): 443–61 (447–51).

41 Wang, *Xunzi Jijie* (note 38), 316.

42 In another article, I argue that Xunzi perceives ritual as a communal technology of the body that allows humans to transform their bodies and minds and obtain physical and spiritual bounties while at the same time enhancing sociopolitical

stability and harmony ("Xunzi's Theory of Ritual Revisited: Reading Ritual as Corporal Technology," *Dao: Journal of Comparative Philosophy* 12, no. 3 [2013]: 313–30).

43 Wang, *Xunzi Jijie* (note 38), 316–17.

44 Pines, "Disputers of the Li" (note 14), 39.

45 Wang, *Xunzi Jijie* (note 38), 318–19.

46 Mu-chou Poo, *In Search of Personal Welfare: A View of Ancient Chinese Religion* (Albany: State University of New York Press, 1998). For more details, see Daniel Sou's contribution to this volume, "Living with Ghosts and Deities in the Qin State," which analyzes the recently excavated daybook [*rishu*, 日書] exorcism manuals from the Shuihudi corpus.

47 Chi-tim Lai, "The Opposition of Celestial-Master Taoism to Popular Cults during the Six Dynasties," *Asia Major* 11, no. 1 (1998): 1–20 (11–13).

48 Marianne Bujard, *Le sacrifice au Ciel dans la Chine ancienne: Théorie et pratique sous les Han occidentaux*, Monographies de l'École française d'Extrême-Orient 187 (Paris, 2000).

49 Constance Cook, *Death in Ancient China: The Tale of One Man's Journey* (Leiden: Brill, 2006); Hung Wu, *The Art of Yellow Springs: Understanding Chinese Tombs* (Honolulu: University of Hawaii Press, 2010).

50 Poo, *In Search of Personal Welfare* (note 46); Puett, *To Become a God* (note 9).

51 Lagerwey and Kalinowski, eds., *Early Chinese Religion: Part One: Shang through Han (1250 BC–220 AD)*, (Leiden: Brill, 2009).

CHAPTER 6

Living with Ghosts and Deities in the Qin 秦 State

Methods of Exorcism from "Jie 詰" in the Shuihudi 睡虎地 Manuscript

Daniel Sou

In early Chinese history, people maintained a close relationship with entities from the nonhuman sphere, including ghosts and deities. Numerous writings and material artifacts from the Shang 商 dynasty to the Han 漢 dynasty reveal that people worshiped supernatural entities, seeking guidance, petitioning for wealth and prosperity, and expressing devotion. No general sentiments describe how the ancient Chinese viewed their relationship with ghost and deities, but they did believe that these unworldly entities had the power to reward and punish behavior. This perspective is well presented in the chapter "Percipient of Ghosts, Part III" (*Ming gui xia*" 明鬼 下) of the *Mozi* 墨子: "Now if all the people of the world believe ghosts and spirits can reward the worthy and punish the wicked, then how could the world be in disorder?"[1] On the evidence of this example, the relationship between humans and nonhuman entities was conditional; immoral and wrongful acts warranted punishment.

Yet the relationship was not exclusively conditional, nor was it simply characterized as a giver-receiver arrangement; in some situations, the relationship turned hostile and frightful, especially when ghosts and spirits chose to inhabit the human realm and haunt the living. What, then, could people living in early China do to protect themselves from harm by an unknown entity? The simple answer is that one had to identify the threat and perform an exorcism.

This essay explains the diverse methods of exorcism practiced by the people of the Qin 秦 state, the kind of supernatural entities they feared, and for what reason. The source of this discussion is a short piece of writing titled "Jie 詰," which is included in *Rishu jia* 日書 甲, or *Daybook Version A*, unearthed from Shuihudi 睡虎地. There are many records about ghosts, deities, and exorcism in various traditionally received manuscripts. Yet "Jie" offers a unique understanding of unworldly entities, detailing exorcism methods and materials that are rarely found in other writings.

Shuihudi Manuscripts, Daybooks, and "Jie"

Shuihudi Manuscripts and Its Daybooks

A proper examination of exorcist practices mentioned in "Jie" requires a basic understanding of the archaeological and textual characteristics of the Shuihudi manuscripts and the daybook in which "Jie" is included. From December 1975 to January 1976, an excavation of tomb M11 in the Qin dynasty cemetery at Shuihudi in Yunmeng 雲夢, Hubei province, unearthed a massive number of bamboo slips. The tomb owner, Xi 喜 [Happy], was probably a low-level scribe in the local government and was buried in 217 BCE, according to the slips, which were found inside his coffin next to the corpse.[2] One important fact is that this region did not originally belong to the state of Qin but to the Chu 楚 state, which was conquered by the Qin in 223 BCE, only seven years before the death of the tomb owner.

The total number of unearthed slips was about 11,000, each one varying from 0.5 to 0.8 centimeters in width and 25.5 to 27.8 centimeters in length, which is about 1 *chi* 尺 (approximately one foot) in length using traditional Chinese measurement. The calligraphy style used on the slips is *lishu* 隸書, or "clerk style," and some strips have writing on both sides.[3] Like all of the excavated manuscripts of this period, we unfortunately cannot say anything definitive about its author(s).

The bamboo slips found inside the coffin consist of ten separate texts, which are *Bian'nian ji* 編年記 [Chronicle], *Yushu* 語書 [Speech Document], *Qin lü shiba zhong* 秦律十八種 [Eighteen Qin Statutes], *Xiao lü* 效律 [Statutes Concerning Checking], *Qin lü za chao* 秦律雜抄 [Miscellaneous

Excerpts from Qin Statutes], *Falü dawen* 法律答問 [Answers to Questions Concerning Qin Statutes], *Fengzhen shi* 封診式 [Models for Sealing and Investigating], *Wei li zhi dao* 為吏之道 [How to Conduct Yourself as an Official], and two versions of daybooks, *Rishu jia zhong* 日書甲種 [Daybook Version A], and *Rishu yi zhong* 日書乙種 [Daybook Version B]. As is usual with many other published excavated slips, all of these titles were given by the editorial team who have transcribed and ordered the slips.[4]

The reason the editorial team titled the last two texts *Rishu*, or *Daybook*, is that this very term was written on Shuihudi manuscript slip 260, the last bamboo slip of what is now entitled *Daybook Version B*. And since versions A and B contain similar contents, such as records of hemerological calendars and day omens regarding mundane life concerns, both received the same title.[5] However, the "Jie" section under consideration in this paper is included only in *Daybook Version A*.

Certainly, there are other daybooks excavated from different regions of other states and dynasties. Examples include 1) the two sets of Qin daybooks found in tomb M1 of Fangmatan 放馬灘, Tianshui 天水 city of Gansu province, dating back to the late third century BCE; 2) a Chu 楚 state daybook, probably the earliest example of its kind, from Jiudian 九店, Jiangling 江陵 county of Hubei province; and 3) a Han dynasty daybook from tomb M8 of Kongjiapo 孔家坡, Suizhou 隨州 of Hubei province. All of these daybooks feature hemerological calendars and daily omens regarding mundane activities for ordinary people and officials, such as traveling, catching thieves, and arranging official meetings.[6] But none of them have a section entitled "Jie" or instructions for exorcism.

Unfortunately, we do not know why there is only one instance of "Jie" among the excavated daybooks. More knowledge in this area would help measure the popularity of certain exorcism methods in early China proper. More than likely, because each daybook is distinct in some way, whether big or small, they all show some degree of customization.[7] There are even differences between the two Qin daybooks from Shuihudi and Fangmatan, suggesting that the authors might have changed some content according to their own interests; some writers probably viewed "Jie" as important and interesting, while others did not. Today, scholars must simply wait for another excavated daybook to seek answers.

A Brief Survey of "Jie"

The "Jie" section of *Daybook Version A* identifies seventy cases of various types of harm done by ghosts, deities, and demonic creatures, as well as by natural yet unusual phenomena. Among these seventy cases are forty records of ghosts; eight natural phenomena, such as thunder or unusual cold; and six demonic creatures. Apparently, ghosts were the greatest concern of the people. Every entry about ghosts is written in a formal way and starts with a brief explanation of the harm done, the name of the ghost or deity that caused it, how to prevent or expel the harm through an exorcism, and, often, the result of the exorcism. Thus, the chapter as a whole can be regarded as a manual for the practice of exorcism, a certain kind of demonography authorized by the Qin state.[8]

Within the seventy entries in the "Jie," most of the ghosts and deities are anthropomorphic beings, and little distinction is made between the two in terms of the harm they cause. The only noticeable difference between the two types of entities can be found in one example about the "Upper Deity" (*shangshen* 上神), who flirts with young boys and girls (*Shuihudi Qin mu zhujian* 1990: slip 31.2 [hereafter *Shuihudi*])[9] or demands someone's wife (*Shuihudi* 1990: slip 39.3). In both instances, the author uses the character *xia* 下, or "coming down," which indicates that the Upper Deity came down from the upper realm. No such notion is expressed regarding ghosts. However, considering the types of harm that the two caused and their outward guise, there is little distinction recorded.

Although the chapter provides information about the harm done by ghosts and deities, in many cases, the author indicates that the harm happened for no apparent reason. The phrase "with no reason" (*wugu* 毋故) appears eighteen times among the seventy entries, indicating that people experienced all sorts of harassment or harm by ghosts and deities without explanation. For example, a "stabbing ghost" (*cigui* 刺鬼) would continually attack people (*Shuihudi* 1990: slip 27.1) and a "mound ghost" (*qiugui* 丘鬼) would stay in one's house (*Shuihudi* 1990: slip 29.1), both without any apparent reason. Generally speaking, people believed that harm was payback for wrongdoing, that ghosts and deities punished those who wronged them in life and helped those who did favors for them.[10] However, in the "Jie" section, there is no mention of any retribution or reward from ghosts and deities, and according to the author(s) of "Jie," ghosts were just evil and wicked, harming people out of spite alone: "Ghosts harm people, behave wantonly, and do inauspicious things to people" (*Shuihudi* 1990: slip 24.1).

The nature of these "troublemakers" and the methods to repel them are grounded in the title, "Jie," which is noted on top of *Shuihudi* manuscript slip 24.1. Translating the chapter title correctly is the first step in analyzing the chapter because the title epitomizes the content. In previous studies, there are at least three possible translations:

1. "Inquiry": The chapter provides "instructions for handling various evil ghosts, spirits, and even gods."[11]
2. "Spellbinding": Although the word *jie* refers to a legal term meaning "interrogation," it "still applied to older practices in which oaths and spells had the power to magically obligate men and demons."[12]
3. "Prevention": Based on *Zhou li*'s 周禮 [Rites of Zhou] definition of the word, and other texts listed in "Yiwen zhi 藝文志" [Treatise on Arts and Literature] of *Hanshu* 漢書 [Han Records] that use the word *jie*, it means "to prevent," especially to prevent harm from ghost, deities, and other supernatural creatures.[13]

In fact, all three translations are acceptable because each represents the contents of the "Jie" section. Using "inquiry" or "interrogation" is acceptable because the chapter's written form resembles a legal document and because this chapter (and the manuscript that includes the chapter) was in the hands of the local government and eventually placed in the tomb of a local judicial officer. "Spellbinding" is somewhat problematic. Although the chapter does reveal methods for commanding ghosts and deities to retreat and to prevent harm, there is no exorcism based on chanting (casting spells). Instead, requests appear for people to perform a certain rite or action using a certain material. Liu's translation, "prevention" or "to prevent," suits the literal and etymological definition of *jie* but again fails to uncover any religious or legal implications of the chapter. Based on this analysis, there seems to be no appropriate way to translate *jie* using one word. Thus, it is important to acknowledge that the title essentially means "to prevent" harm by ghosts and deities using something akin to "magical powers"; nevertheless, the chapter is written in a formal tone appropriate to the local government.

Human Ghosts

In many cases in early China, ghosts were perceived as entities with concrete forms. In other words, they had visible physical features with definite shape, sometimes human. During the Warring States period, Xunzi 荀子 noted that the reason to place burial deeds inside the tomb was to help the ghost live like a human being.[14] This logic is based on the idea that a ghost is what a human transforms into after death, as mentioned in the *Mozi*: "The ghosts and spirits of ancient and modern times are the same. There are ghosts of Heaven, ghosts and spirits of the mountains and rivers, and ghosts of people who have died."[15] Later, in the late Han dynasty, Wang Chong 王充 argued for the commonly held belief of his time that ghosts looked like human beings. Based on his argument, the Han people clearly thought that all ghosts not only possess the form of human beings but also live like human beings: eat, drink, and even think as humans do.[16] All of these statements imply that ghosts—not all, but some—were once humans and reappear in their previous shape; hence, I refer to them as "human ghosts." In keeping with these three received texts, the "Jie" section contains records of four human ghosts:

> If someone's wife, consort, or friend dies and their ghost returns, boil cyperi rhizoma and thorn wood branch [together] and give this mixture to the ghost; then it will stop coming (*Shuihudi* 1990: slips 65.1–66.1).

人妻妾若朋友死，其鬼歸之者，以莎芾牡棘枋(柄)，熱(爇)以寺(待)之，則不來矣。

For no apparent reason, everyone in the house gets smallpox. Some die and some get sick. Men and women lose their hair and have yellow pupils. This is caused by a person swaddled in cloth who became a ghost (i.e., an infant ghost). Use one *sheng* 升 of villous amomum fruit to fill the pounding mortar, [grind it with] millet and meat to feed the person (the infant ghost) who starved to death; then it will stop. (*Shuihudi* 1990: slips 43.1–46.1)

人毋(無)故一室人皆疫，或死或病，丈夫女子隋(墮)須(鬚)羸髮黃目，是宲宲(是是宲)人生為鬼，以沙人一升[扌/室]其舂臼，以黍肉食宲人，則止矣。

A ghost who always comes into a house naked is a child who died from injury and was not buried [properly]. Spread ashes on the bones; then it will stop coming. (*Shuihudi* 1990: 50.2)

鬼恒羸(裸)入人宮，是幼殤死不葬，以灰濆之，則不來矣。

A baby ghost who always shouts at people saying, "Give me food!" is a sorrowful breast-feeding ghost. Cover the bones that are exposed with yellow dirt; then it will stop [asking for food]. (*Shuihudi* 1990: 29.3–30.3)

鬼嬰兒恒為人號曰:「鼠(予)我食」。是哀乳之鬼。其骨有在外者，以黃土濆之，則已。

Among the forty records of ghosts, the four listed above are related to human ghosts, that is, ghosts that were once human. The first case shows how to remove a human ghost, supposedly an adult when alive, using cyperi rhizome and a thorn-wood branch. Including this entry, five records refer to thorn wood as a material from which to make exorcising weapons, such as arrows (*Shuihudi* 1990: slips 28.1 and 66.1) and swords (*Shuihudi* 1990: slips 42.2, 25.3, and 36.3). These records imply that people believed in thorn wood's exorcising power. Not only in the "Jie" chapter, but also in the

Chuqiu Zuo zhuan 春秋左傳 [Zuo commentary on the Spring and Autumn Annals; hereafter *Zuo zhuan*], we can see that the thorn-wood arrow was used as an exorcising weapon: "A peach tree bow and thorn wood arrow will prevent disaster."[17]

Interestingly, among the four human ghosts, three are infant ghosts who haunt people because they were improperly buried or died of starvation. The suggestion of offering millet and meat, which were probably expensive for many at that time, to the baby ghost implies that the baby ghost is haunting the people because it died of starvation. The disease that causes one to lose hair and have yellowish pupils is unidentified in this section and throughout the Shuihudi manuscripts, but such symptoms were known to be signs of malnutrition. In revenge for dying of starvation, the baby ghost harms people, most likely family members, by making them suffer with the symptoms of malnutrition.

The other two baby ghosts haunt people because of the babies' wrongful burial. In response to the baby ghost's request for food in Shuihudi slips 29.3 through 30.3, the passage suggests using ashes and yellow dirt to cover the exposed bones of the dead. These instructions note that ghosts were unsettled by an improper burial or because their bones were exposed. Stories of improperly buried individuals harming people became widespread after the Han dynasty and are often seen in the *zhiguai* 志怪, or "Accounts of the Strange" genre, in which ghosts request that someone, usually not family members but rather a stranger or a government official, carry out a reburial and then in return provide a reward for his or her generosity.[18]

Another reason for people's concern about exposed skeletons was that they knew that an exposed corpse could cause plague. Therefore, the duty of an official was to search for exposed corpses and rebury them or ask wealthy families to help the poor conduct a proper burial.[19] Whether a religious belief or a hygienic issue, as the *Li ji* 禮記 [Book of Rites] suggests, ghosts had to be buried, and their bones and flesh had to decompose under the earth.[20] These rules were common among people from the Qin and the Han and perhaps emerged earlier than the Qin.

The simple reason for using yellow earth and ashes for exorcism might be that covering the exposed corpse in this way symbolized a reburial, thus reassuring infant ghosts. But why did they use these two substances and nothing else? And what kind of ashes? Although there is no answer provided in the Shuihudi manuscripts, ashes and yellow earth might be the same material, based on the *Wushier bingfang* 五十二病方 [The Recipes for Fifty-Two Ailments] in the Mawangdui 馬王堆 silk manuscript. This text instructed those who suffered from *baichu* 白處, probably an illness featuring white patches on the skin, to take one tenth of a *sheng* 升 of "yellow earth from the stove," mix it with azurite and salt, smooth it on the skin, and drink it until the symptoms faded.[21] Further, "stove residue ashes" mixed with water were given to those who had been bitten by a mad dog.[22] As Donald Harper correctly pointed out in *Early Chinese Medical Literature*, the yellow earth from the stove was no different from the stove residue, which has a medicinal effect. Other than the Mawangdui manuscript, there is no single reference to yellow earth and stove ashes in a single context or any mention of their medical use. Although the Mawangdui manuscript restricted use of the two substances to physical treatment, the "Jie" section might have extended such healing features to the spiritual sphere.

Moreover, the three records of baby ghosts also might imply the high death rate of infants during that period caused by a shortage of resources and might indicate that family members usually did not provide a proper burial for infants or children. A lack of money may have been one reason that people did not provide a proper burial. However, the expenditure for an infant's burial probably would not have been high; thus, there might have been another reason for the neglect. People did not consider babies and young children to be fully functional human beings. According to Robin Yates's research, an "employable" child was seven to fourteen years old, while a "not yet employable" child, an infant, was six or younger. In this context, "employable" means those who are registered and mainly work and serve in the military, which every Qin citizen was obligated to do into his or her sixties.[23] So it is possible that people did not concern themselves much with the death of young, "unemployable" children and neglected to give them a proper burial.

Licentious Deities and Demonic Creatures

While reading the "Jie" section, one finds an interesting theme that is not found in either the pre-Han received texts or newly excavated bamboo manuscripts: sexual harassment. Not to be confused with later *zhiguai* [Accounts of the Strange] stories about love affairs between ghosts and humans, mostly males, the entries in the "Jie" have nothing to do with "romance" or "revenge" owing to a lovers' betrayal; instead, they reveal both physical and sexual violence done by ghosts and demonic creatures.

The way that licentious ghosts or deities harass people resembles that of a vicious human, implying that these entities were personified in forms and character:

> When a person with the form of a bird, a beast, or [one of] the six domestic animals often intrudes on someone's house, this is the Upper Deity who likes to come down and enjoys having intercourse with virgin boys and girls.[24] Beat drums and shake wood clappers to make a [loud] sound; then it will stop coming. (*Shuihudi* 1990: slips 31.2–33.2)
> 人若鳥獸及六畜恒行人宮，是上神相好下，樂入男女未入宮者，毄(擊)鼓奮鐸喿(譟)之，則不來矣。

Although in this case, a deity, not a wicked ghost, comes down from the upper realm to have sexual intercourse with youngsters, confirming a physical relationship between godlike figures and human beings, this behavior is similarly viewed as harassment. With regard to exorcism, the record suggests making loud sounds with a drum and wooden clappers. As with many other examples in the "Jie" section, the reason for beating drums is unrevealed. Another example of the use of drums in the "Jie" is when the people are instructed to beat a drum when a ghost is beating a drum inside one's house: "There is a sound of drumming inside the room, [but you] cannot see the drum; this is [caused by] a ghost drumming. Respond to the ghost with a human-made drum; then it will stop" (*Shuihudi* 1990: slip 34.3).

Indeed, beating a drum was commonly known as an exorcising act, even before the Qin dynasty. According to the *Zuo zhuan*, drums were used whenever there was an eclipse or flood: "[In the twenty-fifth year of Duke Zhang], in the sixth month, *xinwei* 辛未, the first day, there was an eclipse. [So] beat the drum and offer a sacrifice at the shrine" (*Chunqiu Zuo zhuan*: 10/320).[25] Although the *Zuo zhuan* lacks explanation for using a drum, the *Chunqiu Gongyang zhuan* 春秋公羊傳 [Gong Yang commentary on the Springs and Autumns Annals] suggests that people beat drums to give more *yang* 陽 energy to the declining sun, which is dominated by *yin* 陰 energy (*Chunqiu Gongyang zhuan* 8/200). The sound of drums likely represents the sound of Heaven, according to Xunzi, which makes the drum the most powerful instrument of all (*Xunzi jijie* 20/383). Perhaps the low, deep sound resembles the sound of thunder and thus represents rage, evoking fear in the otherworldly entities.

Another licentious unnamed entity demands someone's daughter to take as its bride or consort: "If a ghost always says to someone, 'Give me your daughter,' and you cannot refuse, this is a Higher Deity coming down to have a wife. Beat it with reed; then it will die. [If you] do not protect [your daughter], the Higher Deity will come five times and [eventually] the girl will die" (*Shuihudi* 1990: slips 39.3–40.3). The interesting point here is that this ghost calls itself the "Upper Deity," the same deity as noted above.

The method of using a reed to repel harassing entities is described in the case of the "son of the Higher Deity."

> A ghost always follows a woman and stays by her side, saying, "The son of the Higher Deity came down to have fun." [If you] want to get rid of it, wash yourself with dog excrement and beat it with reed; then it will stop. (*Shuihudi* 1990: slip 38.3)
> 鬼恒從人女，與居，曰：「上帝子下游。」欲去，自浴以犬矢，毄(繫)以葦，則死矣。

A famous record using dog excrement for exorcism is also found in the *Hanfeizi* 韓非子. Here the wife of Li Ji 李季 of the Yan 燕 state was having an affair when Li Ji unexpectedly returned home. At that moment, a house servant advised the frightened wife:

> "Ask the young nobleman to get naked and loosen his hair, and rush right away to the gate. We will pretend not to see him." So the young nobleman followed the servant's advice and ran quickly through the gate. Li Ji asked, "Who is that?" Everyone said, "There was nothing." Li Ji said, "Did I see a ghost?" The wife said, "Yes." Li Ji said, "Then what I shall do?" The wife said, "Collect excrements from the five animals and wash yourself." And Li Ji said, "Okay."[26]
> 其室婦曰: "令公子裸而解髮, 直出門, 吾屬佯不見也。" 於是公子從其計, 疾走出門。季曰: "是何人也?" 家室皆曰: "無有。" 季曰: "吾見鬼乎?" 婦人曰: "然。" "爲之奈何?" 曰: "取五牲之矢浴之。" 季曰: "諾。"

According to Wang Xianshen's 王先慎 commentary on this story, the "five animals" are an ox, sheep, pig, dog, and chicken. Although there is a difference in terms of which animals are the source of excrement, both the "Jie" and the *Hanfeizi* agree that bathing in excrement is a method for removing a ghost.

Other records that note the use of excrement for exorcism in the "Jie" includes throwing pills made with dog excrement at the Great Deity (*dashen* 大神) who harms people (*Shuihudi* 1990: slips 27–28.2) or at one's ancestor's ghost who peeks inside the house (*Shuihudi* 1990: slip 49.2), or burning pig excrement to get rid of the Yang ghost (*yang'gui* 陽鬼), which takes away the fire for cooking (*Shuihudi* 1990: slip 55.1), or to prevent drooling caused by a ghost named Yuanmu 爰母 (*Shuihudi* 1990: slips 50–51.3). Although no records specify why animal excrement is an effective way to expel ghosts or deities, Liu Lexian suggests that if ghosts and human beings resemble each other in form and character, it is plausible to assume that ghosts hate excrement simply because people detest it.[27]

In addition to ghosts and deities, there were other demonic creatures that would sexually harass people.

> A dog that comes into people's house often at night, forcing down the husband and harassing the wife, and cannot be caught, is a demonic dog disguised as a ghost. Make □□ (unidentifiable graphs) with mulberry tree bark, steam it, and eat it; then it will stop. (*Shuihudi* 1990: slips 47.1–49.1)
> 犬恒夜入人室，執丈夫，戲女子，不可得也，是神狗偽為鬼。以桑皮為□□之，□(炮)而食之，則止矣。

> A ghost that often follows men and women, and will disappear when it sees other people, is a demonic insect disguised as a human being. Cut its head with a fine sword; then it will stop coming down. (*Shuihudi* 1990: slips 34.2–35.2)
> 鬼恒從男女，見它人而去，是神蟲偽為人，以良劍刺其頸，則不來矣。

According to the above examples, both demonic creatures had to mutate from their original form, dog and insect, to a ghost or human to harass people sexually. This entry suggests that a human form was needed for such a heinous act (i.e., the demonic creatures could not have intercourse in their original form). Another possibility is that the two entries reflect a social taboo.

Otherworldly Entities Living in Houses

Ghosts and deities could show up anywhere and do harm to people, but there was one place they were particularly unwelcome: residential areas, particularly inside a house. As noted in the *Lunyu* 論語, "Revere ghosts and deities but keep them at a distance; this is what we say is wise," indicating that although Confucius acknowledged the power of ghosts and deities, they were still fearsome entities that should be kept at a distanced.[28]

Among more than seventy entries, a total of eighteen cases concern the harm done by ghosts or demonic creatures that stay inside one's house and cause various harm to residents. The cases are categorized by name, harm, and method of exorcism in the table below.

Table 6.1

Name	Harm	Exorcism	Slip no.
Mound ghost (丘鬼)	None described	Make human and dog figures, place them around one's house, throw ashes to the ghost, and hit the window, making a loud sound.	29–30.1
Thorny wood ghost (棘鬼)	Plague (疫)	Dig out the ghost staying underneath the house and get rid of it.	37–39.1
Pregnant ghost (孕鬼)	Plague (疫), dreams, die while asleep	Dig out the ghost staying underneath the house and get rid of it.	40–42.1
Young ant (幼蠪)	House is cold during the hot summer	Remove it from one's house.	51.1
Yang ghost (陽鬼)	Weakening kitchen Fire	Burn swine excretion inside the house.	55.1
Yin ghost (欲(陰)鬼)	Death of the six domestic animals	Broken slips.	56–57.1
Harmful ghost (狀(傷)神)	Difficulty of breathing and movement	Cook underground water, a red pig, tail of a horse, and a head of a dog and eat it.	36–38.2

Name	Harm	Exorcism	Slip no.
Five color insect (會蟲)	Muscle contraction	Rip off five chi of earth from the southwestern corner of the room, and pick out the insect with an iron stick. Then hit the middle of the insect's head and dig a hole and get rid of it.	39–41.1
Wandering ghost (游鬼)	Coming inside house	Make a black kite wing with guang'guan (廣灌) grass and burn it.	51.2
Unfortunate ghost (不辜鬼)	Death of infants	Scatter ashes on the door at day geng 庚, when the sun rises, and perform a ritual for ten days. Wrap the ashes with white cogon grass (imperata cylindrica, 白茅) and bury it in the field.	52–53.3
Canya ghost (粲迓之鬼)	Injury inside house	Mix white cogon grass (白茅) with yellow earth and clean the room with it.	57–58.2
Qi of acarid (恙氣)	Hair curls like an insect antenna	Broken slips.	60–61.2

continued

Table 6.1 (continued)

Name	Harm	Exorcism	Slip no.
□鬼 (unable to identify)	Nightmares	Hit the four corners and the middle of the room with a peach wood broom, cut the walls with a thorn wood sword, and shout, "Quick, run out! If you don't go out today, I will cut off your skin with this sword and wear it as clothes."	24–26.3
Fearful ghost (遽鬼)	Coming inside house	Throw rocks at the ghost.	28.3
Dixue (地辥)	People falling down while asleep	None described.	31.1
Ghost drumming (鬼鼓)	Drumming inside House	Respond with a human-made drum.	34.3
Yuanmu (爰母)	People dribbling	Dig a three *chi* deep hole and burn pig excretion inside it.	50–51.3
Small pox ghost (癘鬼)	Itchy skin	Burn fresh paulownia tree inside the house.	52.3
Unnamed insects	Bloody well water	Eat steamed rice and drink dew.	53–56.3

Figures and Dogs for Exorcism

The meaning of making human and dog figures of dirt to repel a mound ghost (*Shuihudi* 1990: slips 29–30.1) can be understood by analogy with the ancient use of clay figurines and the symbolic representation of a dog. To prevent disasters, *Lunheng* instructs people to make "ghost-like" (*guixing* 鬼形) clay figures for a ritual to the "Earth Deity" (*tushen* 土神), and the *Wushier bingfang* notes that a method to remove a child ghost involved fastening one figurine above each side of the doorway.[29]

According to the "Jie" section, dogs seem to have had exorcistic power; for example, a dog figure could repel harm from a ghost (*Shuihudi* 1990: slips 29–30.1), and eating a dog's head, mixed with other ingredients, could remedy a breathing problem (*Shuihudi* 1990: 36–38.2). Another example comes from a spring Nuo 儺 sacrifice found in the *Fengsu tongyi* 風俗通義 [Comprehensive meaning of customs]; people were instructed to dismember dogs at the township's four gates in order to complete the *qi* 氣 of spring.[30] Dogs were sacrificed because they represented metal (*jin* 金) according to the five-phase theory, followed by wood (*mu* 木), which governs.[31] When metal was blocking wood, preventing spring from arriving, then the ritual was to sacrifice the representative animal, dog for metal.

Plants for Exorcism

Plants, such as peach trees, white cogon grass (*imperata cylindrica)*, and paulownia trees, also were believed to have exorcising power. As explained in the chart above, people made a peach-wood bow and a thorn-wood arrow to prevent disaster, and here it is a "peach-wood broom" (*Shuihudi* 1990: slips 24–26.3). Using both a bow and a broom implies that the object made of peach wood was not the main concern but rather the peach wood itself.[32]

White cogon grass (*baimao* 白茅) or *Imperata cylindrica* (*Shuihudi* 1990: slips 52–53.3; 57–58.2) often was used as a receptacle, a material in the place of animal sacrifice, or a material used in burial rites or burial goods. For example, *Shiji* 史記 [Records of the Grand Scribe] notes that each feudal lord of the five regions would wrap his earth color with white cogon grass, take it back to his region, and build an altar on top of it.[33] The function of white cogon grass is described in Shuihudi manuscript slips 52 to 53.3 as an exorcist material to repel an "unfortunate ghost" causing the death of infants, and as a receptacle in which to wrap the exorcising material: ashes or the five colors of earth. The Mawangdui *Wushier bingfang* also notes that white cogon grass was a replacement for a pig sacrifice when the practitioner was loath to kill a pig, thinking the sacrifice inhumane.[34] Another example is in the Fangmatan 放馬灘 bamboo manuscript: "People of the market [i.e., ordinary people] think white cogon grass is valuable, and the ghost takes it and [also] thinks it is valuable" 市人以白茅為富，其鬼受於他而富.[35] White cogon grass, like peach wood, was regarded as a religious material and was used in various shapes and ways. As for paulownia wood, both *Yantie lun* 鹽鐵論 [Discourses on Salt and Iron] and *Chunqiu fanlu* 春秋繁露 [Luxuriant Gems of the Spring and Autumn Annals] record it as a material in which to carve a horse (along with a human clay figure) or a fish figure to be used in rituals.[36]

The three cases described here are far from providing a solid reason why such plants were used in rituals or possessed exorcising power more than did others. However, it is significant that in all cases these writings claimed these plants had the power to prevent harm from spiritual beings.

Possibility of Governmental Supervision

After reading the "Jie" section and trying to place it among the other sections of the Shuihudi manuscripts, one question emerges: why are recorded exorcisms part of an official document? Actually, this question also applies to the two sets of daybooks found along with the excavated Qin legal documents from Shuihudi. As noted by Mark Lewis, the daybooks were probably used for political guidance by local officials and carried equal or similar authority to Qin law: "Because these mantic texts (i.e. the two daybooks) were buried together with the legal materials, it is likely that the deceased official (i.e. 'Xi') or his subordinates employed them in their everyday administrative activities, further blurring the line between legal and religious practice."[37]

The daybooks provide profound information about early Chinese religious belief and activities, but for their contemporaries—a Qin official, like the tomb owner Xi—these daybooks, or "mantic texts," had political import as well. What kind of political aims did daybooks serve? I believe there can be two answers in regard to the "Jie" section.

First, local government could establish social order by providing methods of exorcism to those who were haunted. None of the entries in "Jie" mentions a "specialist": a mediator between the human and nonhuman world, a shaman who could perform exorcism, also known as an oneirocritic. As noted in the first section of this essay and as seen in various entries, the one who was haunted performed the exorcism. In other words, the entire chapter might be regarded as a kind of do-it-yourself manual for exorcism; a collection of exorcisms written and preserved by a local official. Thus, the local government might have been responsible for providing information about exorcism to the people to help them solve their problems, promoting social stability and order.

Second, the "Jie" section likely helped control local customs, including religious activities, in order to preclude any potential threat to the Qin authority. After all, the region in which archaeologists found the Shuihudi manuscripts belonged to the Chu state, occupied by the Qin:

> In ancient times, people had local customs. What was [considered] beneficial and liked or disliked were not the same among different groups. Some customs did not suit the people and could harm the state. Hence, the sage kings developed standards (or laws) and regulations, instead of unreliable and even harmful customs, and to get rid of vice [as well as] their heinous customs. (*Shuihudi* 1990: slips 1–2)
> 古者，民各有鄉俗，其所利及好惡不同，或不便於民，害於邦。是以聖王作爲法度，以矯端　民心，去其邪避(僻)，除其惡俗。

The above passage from the *Speech Document*, a writing from the Shuihudi manuscript, indicates the Qin government banned any customs considered harmful to the state and defined them as "heinous." Here the example notes that thus the "sage kings developed standards (or laws) and regulations," but in reality the Qin government already had one.

According to *Answer to Questions Concerning the Qin Statutes* in the Shuihudi manuscripts, one who performed an "irregular ritual" (*qici* 奇祠), a ritual unauthorized by the Qin government, would be fined two suits of armor (*Shuihudi* 1990: slip 131), underscoring the government's authority over religious activities. If the Qin government strictly supervised rituals and prohibited performing an unauthorized ritual by law, then it probably had laws restricting exorcism. Furthermore, if the Qin government presumed the recorded exorcism and its associated religious beliefs were a threat to the state, there would have been no need to keep a record nor any reason to allow this record to be placed inside the tomb of an official.

A different understanding of Shuihudi daybooks is provided by Kudō Motoo. Based on the *Shangjun shu* 商君書 [Book of Lord Shang], which mentions that a sage should observe current customs in order to establish law and bring order, Kudō suggested that daybooks were documented because the Qin government wanted to know the customs that existed among people. In the case of the Shuihudi manuscripts, these customs belonged to the occupied Chu state.[38] Although his argument is slightly different from what I have presented above, documentation of daybooks, including "Jie," was politically expedient for the Qin government.

Conclusion

People who lived in the late Warring States period believed in spiritual or supernatural beings and held a strong belief that there was a way to remove them from their daily lives. Many entries in the "Jie" chapter reflect the common religious belief that ghosts and deities were anthropomorphic or zoomorphic beings in form and sometimes in character, but differ in regard to whether people believed in any *du et des* (give and take) relationship with those entities.

With no give-and-take relationship with the ghosts and deities, and having realized that they would harm people no matter what, the people had no reason to show them respect. If they did harm at will, people did not have to look at themselves and search for any wrongdoing but rather could concentrate on getting rid of the ghosts and deities. These ghosts and deities were still tightly connected to daily life, but their relationship was neither superior-inferior nor giver-receiver in nature. Of course, there was no way to prevent such harm or to harm ghosts and deities. So people could only fear their appearance. Nevertheless, there was a manual provided by the local government. People did not have to hire a specialist to perform the exorcism, many exorcism materials were easily obtained, and most of all the manual vouched for its results. Thus, the people's attitude toward the supernatural and harmful beings was somewhat complicated. On the one hand, they feared them for causing harms such as diseases, sexual harassment, and death; on the other hand, they knew how to nullify their presence. The "Jie" chapter describes a way for ordinary people to match the power of supernatural beings.

Although the people knew how to practice exorcism and did not need others' help, possibly the local government provided them with support. The local government gathered information, compiled it, and wrote it down as well as controlled the information itself. Without the government or the local officials, the local people might not have been able to overcome the harm or might have taken a great deal of time to figure out how to deal with these ghosts and deities. In truth, this government support was a double-edged sword. Controlling the local customs meant that the government could and would eliminate disagreeable customs that it believed would harm their authority, but as described above, the collected information provided effective solutions to those who needed them. Whatever "edge" one chooses, popular religion cannot be fully understood without considering its sociopolitical context.

So what is the most important point of the "Jie" section? The chapter reveals information about what the people believed about ghosts and deities along with specific methods of exorcism. I admit that there are many things we cannot know, such as why the people used a certain material for exorcism. However, the "Jie" section and the two versions of daybooks turn our attention from elite culture to ordinary people, helping us see a more complete picture of religious activities during the Warring States period and the possibility of state control over religion.

Notes

1 Sun Yirang 孫詒讓, ed., *Mozi xiangu* 墨子閒詁, in *Zhuzi jicheng* 諸子集成, vol. 4 (Beijing: Zhonghua shuju chubanshe, 1996), 31/20.

2 For the initial yet brief excavation report, see Xiaogan diqu di er qi yigong yinong wenwu kaogu xunlian ban 孝感地區第二期亦工亦農文物考古訓練班, "Hubei Yunmeng Shuihudi shiyi hao mu Qin mu fajue jianbao 湖北雲夢睡虎地十一號墓發掘簡報," *Wenwu*, no. 6 (1976): 1–10. Later, a full excavation report—including details about the tomb style of M11, its burial goods, and the tomb owner; an initial, unpunctuated transcription; and photographs of the excavated bamboo slips—was provided in Yunmeng Shuihudi Qin mu bianxiezu 雲夢睡虎地秦墓編寫組, eds., *Yunmeng Shuihudi Qin mu* 雲夢睡虎地秦墓 (Beijing: Wenwu chubanshe, 1981).

3 *Wenwu* (note 2), 1–10.

4 The title names and separation of these ten manuscripts were determined by and follow the translation of the editors of the *Yunmeng Shuihudi Qin mu*, 12–14, except *How to Conduct Yourself as an Official* and *Daybook Version A and B*; for those names and translations, see A. F. P. Hulsewé, *Remnants of Ch'in Law: An Annotated Translation of the Ch'in Legal and Administrative Rules of the 3rd Century B.C., Discovered in Yün-meng Prefecture, Hu-pei Province, in 1975*, Sinica Leidensia (Leiden: Brill, 1985).

5 Shuihudi Qin mu zhujian zhengli xiaozu 睡虎地秦墓竹簡整理小組, *Shuihudi Qin mu zhujian* 睡虎地秦墓竹簡 (Beijing: Wenwu chubanshe, 1990), 179.

6 For further explanation of excavated daybooks, see Ethan R. Harkness, *Cosmology and the Quotidian: Day Books in Early China* (Ph.D. diss., University of Chicago, 2011), 11–47; Kudō Motoo 工藤元男, *Suikochi Shinkan yori mita Shindai no kokka to shakai* 睡虎地秦簡よりみた 秦代の 國家と社會 (Tokyo: Sōbunsha, 1998), 150–59. Although there are many other versions of excavated manuscripts entitled *rishu*, or *daybook*, by modern editors, whether this term is a genre distinction is still a question. Harkness proposed five characteristics of a "standard" daybook: presence of hemerological calendars, known as *jianchu* 健除 [Establishment and Removal] or *congchen* 叢辰 [Thicket-of-Branches]; passages concerning an as-

trological entity, such as Taisui 太歲, Dashi 大時, or Xianchi 咸池; passages classifying days using gender-specific values; passages about taboo days based on stem or branch days, often including diagrams; and passages about taboo days related to various mundane topics, such as birth, construction, illness, and catching thieves. He considers daybooks that do not contain these five features to be "non-standard." See Harkness, "Cosmology and the Quotidian," 12. From my point of view, the categorization of "standard" and "non-standard" does not automatically indicate a "genre."

7 Harkness, "Cosmology and the Quotidian" (note 6), 106–7.

8 Donald Harper, "A Chinese Demonography of the Third Century B.C.," *Harvard Journal of Asiatic Studies* 45, no. 2 (1985): 459–98 (460).

9 All of the quoted Shuihudi slips are based on the *Shuihudi Qin mu zhujian* 睡虎地秦墓竹簡 [Bamboo slips from a Qin tomb at Shuihudi], published in 1990 (note 5). Moreover, I have only noted the slip numbers, not the page numbers from the monograph. This monograph, the *Shuihudi Qin mu zhujian*, transcribed the original graphs into modern Chinese and features full annotation, photographs, and translations of the entire set, except for the daybook slips (*rishu* 日書). Before this source, there were several others, such as the *Yunmeng Qin jian shiwen* 雲夢秦簡釋文 in 1976, *Shuihudi Qin mu zhujian* 睡虎地秦墓竹簡 in 1977, *Shuihudi Qin mu zhujian* 睡虎地秦墓竹簡 in 1978 (same title but different text), and *Yunmeng Shuihudi Qin mu* 雲夢睡虎地秦墓 in 1981. But these four publications are less reliable than the 1990 publication for the following various reasons: daybooks were excluded from the 1976 and 1978 versions, the transcriptions are in simplified Chinese, and the annotations are incomplete.

10 Mark Edward Lewis, *The Early Chinese Empires: Qin and Han* (Cambridge, MA: Belknap Press, 2007), 194–95; Michael Loewe, *Divination, Mythology and Monarchy in Han China* (New York: Cambridge University Press, 1994), 39.

11 Poo Mu-chou, *In Search of Personal Welfare: A View of Ancient Chinese Religion* (Albany: State University of New York, 1998), 79.

12 Harper, "A Chinese Demonography" (note 8), 478–79.

13 Liu Lexian 劉樂賢, *Shuihudi Qin jian Rishu yanjiu* 睡虎地秦簡日書研究 (Taipei: Wenjin chubanshe, 1994), 249–50, 265–66.

14 Wang Xianqian 王先謙, ed., *Xunzi jijie* 荀子集解 (Beijing: Zhonghua shuju, 1996), 19/358–9, 19/368–9.

15 *Mozi xiangu* (note 1), 31/153.

16 Huang Hui 黃暉, ed., *Lunheng jiaoshi* 論衡校釋 (Beijing: Zhonghua shuju, 1996), 20/871.

17 Kong Yingda 孔穎達, ed., *Chunqiu zuozhuan zhengyi* 春秋左傳正義. In *Shisan jing zhushu* 十三經注疏, vol. 16 (Beijing: Beijing daxue chubanshe, 2000), 42/1377.

18 Robert F. Campany, "Ghosts Matter: The Culture of Ghosts in Six Dynasties Zhiguai," *Chinese Literature: Essays, Articles, Reviews* 13 (1991): 15–34 (26–27).

19 Mark Edward Lewis, *The Construction of Space in Early China* (Albany: State University of New York Press, 2006), 56–57.

20 Shen Xiaohuan 沈嘯寰, ed., *Li ji jijie* 禮記集解 (Beijing: Zhonghua shuju, 1995), 24/1219.

21 Donald Harper, *Early Chinese Medical Literature: The Mawangdui Medical Manuscripts* (New York: Kegan Paul International, 1998), 247.

22 Ibid., 235.

23 Robin D. S. Yates, "Social Status in the Ch'in: Evidence from the Yün-meng Legal Documents, Part One: Commoners," *Harvard Journal of Asiatic Studies* 47, no. 1 (1987): 197–235 (207–8).

24 Proper punctuation of the second and third lines is debatable. For example, Liu Lexian reads, "是上神相好下，樂入男女未入宮者" (Liu Lexian 劉樂賢, *Shuihudi Qin jian rishu yanjiu* 睡虎地秦簡日書研究 [Taipei: Wenjin chubanshe, 1994], 229); and Wu Xiaoqiang reads, "是上神相好下樂入，男女未入宮者" (Wu Xiaoqiang 吳小強, *Qin jian Rishu jishi* 秦簡日書集釋 [Changsha: Yuelu shushe, 2000], 397). Regardless of which punctuation style is correct, the overall picture is the same: the Upper Deity likes to come down to the human sphere and have intercourse with young boys and girls. The translation problem facing the two scholars above is the graph "*xiang* 相," however one tries to read these two sentences.

25 The same process is recorded in "Duke Wen (文公), year fifteen" of the *Chunqiu Zuo zhuan* (19b/637). According to Michael Loewe, one procedure to end an excessive rainfall was to beat drums for three days along with giving an invocation. For a full translation of the invocation, see Loewe, *Divination, Mythology and Monarchy in Han China* (note 10), 156.

26 Wang Xianshen 王先慎, ed., *Hanfezi jijie* 韓非子集解 (Beijing: Zhonghua shuju, 1998), 31/245–6.

27 Liu Lexian, *Shuihudi Qin jian Rishu yanjiu* (note 13), 263.

28 Liu Baonan 劉寶楠 ed., *Lunyu* zhengyi 論語正義 in *Zhuzi jicheng* 諸子集成, vol. 1 (Beijing: Zhonghua shuju, 1996), 6/22/126.

29 Harper, *Early Chinese Medical Literature* (note 21), 302.

30 See also Derk Bodde, *Festivals in Classical China: New Year and Other Annual Observances During the Han Dynasty, 206 B.C.–A.D. 220* (Princeton: Princeton University Press, 1975), 320–21.

31 Wang Liqi 王利器, ed., *Fengsu tongyi jiaozhu* 風俗通義校注 (Beijing: Zhonghua shuju, 2000), 8/377.

32 Bodde, *Festivals in Classical China* (note 30), 132–34. According to his research, people used the peach tree, its fruit, and its flowers to make a bow or a soup for New Year's Day or a club to kill Archer Yi. In the other two cases, the intent was to prevent harm or disasters by ghosts and spirits.

33 Sima Qian 司馬遷, *Shiji* 史記 (Beijing: Zhonghua shuju, 1982), 30/2115.

34 Harper, *Early Chinese Medical Literature* (note 21), 268–69.

35 Gansu sheng wenwu kaogu yanjiusuo 甘肅省文物考古研究所, ed., *Tianshui Fangmatan Qin jian* 天水放馬灘秦簡 (Beijing: Zhonghua shuju, 2009), slip 5.

36 Wang Liqi 王利器, ed., *Yantie lun jiaozhu* 鹽鐵論校注 (Beijing: Zhonghua shuju, 2000), 29/353; Su Yu 蘇輿, ed., *Chunqiu fanlu yizheng* 春秋繁露義證 (Beijing: Zhonghua shuju, 1992), 74/435.

37 Lewis, *Early Chinese Empires* (note 10), 231.

38 Kudō Motoo, *Suikochi Shinkan yori mita Shindai no kokka to shakai* (note 6), 159.

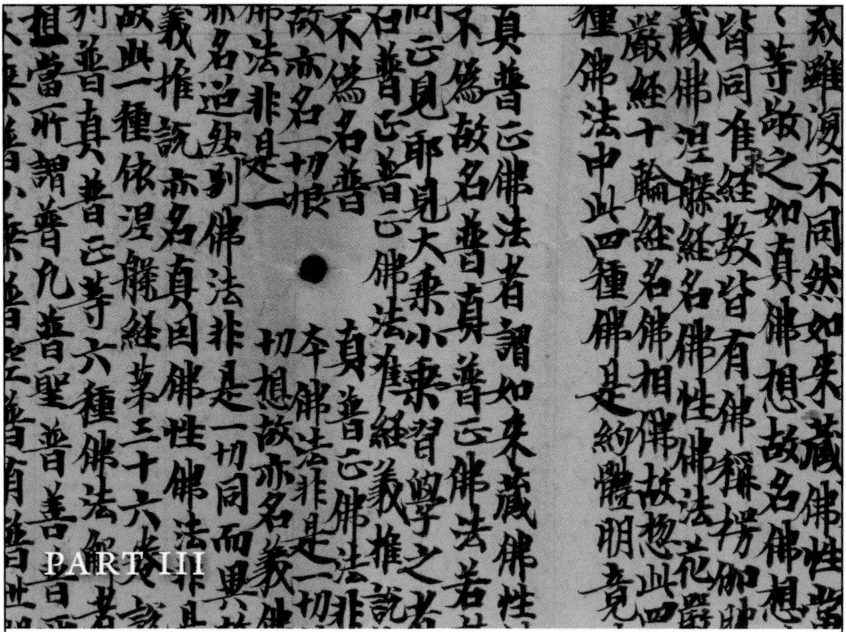

PART III

Technologies of Writing

CHAPTER 7

Spoken Text and Written Symbol

The Use of Layout and Notation in Sanskrit Scientific Manuscripts

Kim Plofker

Because of the traditional reverence for oral composition and recitation in Sanskrit literature, most classical Sanskrit treatises, including scientific ones, were composed in verse and intended (at least in theory) for memorization. Written versions of Sanskrit texts are often presented in imitation of their ideal oral form, as an almost continuous and unformatted stream of syllables. Manuscripts of technical works on subjects, such as mathematics and astronomy, however, had to combine this "one-dimensional" text stream with graphical and notational features generally requiring two-dimensional layout, such as tables, diagrams, and equations. The ways in which this synthesis could be achieved posed several significant challenges for Sanskrit scribes.

In fact, the notion of a Sanskrit scientific manuscript is in some respects almost a contradiction in terms. The following quotation illustrates a long-standing attitude of skepticism toward written texts in the Sanskrit intellectual tradition:

> *pustakasthā tu yā vidyā parahastagataṃ dhanaṃ* |
> *kāryakāle samutpanne na sā vidyā na tad dhanam* ||
> Knowledge which is in a book, money in someone else's hand: when the time comes to use it, that knowledge or that money is not there.[1]

The earliest surviving form of Sanskrit is venerated as a divine speech in which the ancient Indian sacred texts called the Veda were (and still are) recited and transmitted. Most of them are composed in metrical verse, and pandits developed a complicated system of mnemonic cues and memory training to ensure that they were preserved in an oral tradition with every syllable and accent intact. This focus on orality also seems to have inspired the early Indian interest in phonetics and grammar, and the very sophisticated analyses of these subjects dating back to the late first millennium BCE.

At the same time, Indian literature developed a strong dependence on the written word. The traditional oral instruction and explanation that had routinely accompanied the teaching of memorized texts crystallized into written prose commentaries that were disseminated along with, and sometimes instead of, the original verse compositions. The spread of literacy and of intellectual addiction to literacy is thus far familiar in other ancient cultures as well. But the Sanskrit tradition was unique in its persistence in venerating and preserving the ideal of an oral intellectual tradition, even after it had become inseparable from written documents. Sanskrit manuscripts, especially in the sciences, thus represent a paradox: they are a testimony in writing to the supremacy of speech, and this dual nature is reflected in many of their characteristic features.

Some helpful guidance for the modern reader exploring the world of Sanskrit scientific manuscripts can be found in the words of an earlier "Western" scholar confronting these documents in light of his own culturally defined notion of what a book ought to be. Although he lived nearly a thousand years ago, many of his reactions are easy for moderns to empathize with, coming to a large extent from a shared intellectual tradition. He was Abū Rayḥān Muḥammad ibn Aḥmad al-Bīrūnī, the great eleventh-century polymath from Khwarezm in what is now Uzbekistan, who lived many years as part hostage, part protégé of the Ghaznavid sultans at their court in northwestern India. While there, he learned some Sanskrit and learned more of Sanskrit literature in translation, after which he composed a comprehensive work in Arabic called *Kitāb fī Taḥqīq mā li'l-Hind* (Investigation of What Is in India). Many of his remarks on Indian customs and literature touch on the topic of books:

> The scientific books of the Hindus are composed in various favourite metres, by which they intend, considering that the books soon become corrupted by additions and omissions, to preserve them exactly as they are, in order to facilitate their being learned by heart, because they consider as canonical only that which is known by heart, not that which exists in writing. . . .[2]
>
> They do not allow the Veda to be committed to writing, because it is recited according to certain modulations, and they therefore avoid the use of the pen, since it is liable to cause some error, and may occasion an addition or a defect in the written text. . . .[3]
>
> By composing their books in metres they intend to facilitate their being learned by heart, and to prevent people in all questions of science ever recurring to a *written* text, save in a case of bare necessity. . . . They do not want prose compositions, although it is much easier to understand them.[4]

Despite the official preference for memorized texts, the Indic manuscript tradition expanded prodigiously. It constitutes at the present time the largest group of handwritten documents in the world: over thirty million Sanskrit manuscripts were estimated in the late twentieth century to be extant in collections in India and elsewhere.[5] Perhaps as much as one-tenth of the total contains material in the traditional exact sciences, such as mathematics, astronomy, and astrology.

The physical form of these books whose users traditionally sought to avoid books was conservative in its development. Over the period it took the Western manuscript to morph from a parchment or papyrus roll into a codex, the Sanskrit manuscript seems to have persisted in its original form as a collection of unbound pieces of palm leaf or birch bark. As al-Bīrūnī reported, "They bind a book of these leaves together by a cord on which they are arranged, the cord going through all the leaves by a hole in the middle of each. . . . The proper order of the single leaves is marked by numbers. The whole book is wrapped up in a piece of cloth and fastened between two tablets of the same size."[6] The oblong shape (with the horizontal dimension larger than the vertical) was determined by the longitudinal veins of the palm leaves and was conventionally preserved in birch bark and later in paper manuscripts as well.

Al-Bīrūnī commented on other characteristics that he evidently found unfamiliar or noteworthy:

> Indian scribes are careless, and do not take pains to produce correct and well-collated copies. . . . [An author's] book becomes already in the first or second copy so full of faults, that the text appears as something entirely new. . . .[7]
>
> The Hindus begin their books with *Om*, the word of creation. . . .[8]
>
> They write the title of a book at the end of it, not at the beginning.[9]

We cannot test al-Bīrūnī's assertions on Sanskrit manuscripts contemporary with his own experience of them. Owing to the rigors of climate, very few Indian manuscripts predating the past three or four centuries have survived to the present. But many later examples attest to the conservatism of the tradition, as well as to its gradual modification. For example, the manuscript leaf shown in Figure 7.1, a Purāṇa text in the collection of the University of Pennsylvania, omits the central hole for a binding cord but retains the classic oblong shape. It commences not with the sacred syllable *om* but with another standard introductory invocation, *śrīgaṇeśāya namaḥ*, or "Homage to [the deity of beginnings,] Lord Gaṇeśa." After that, it is just a continuous transcription of the verses of the text.

The writing shows the few standard notational marks of the typical Sanskrit manuscript: the folio number in the upper-left and lower-right margins; the verse numbers punctuating the sequence of verses; and (in the first line) the double vertical bars, or *daṇḍas*, that serve as a generic punctuation mark. (The diagonal pattern in the verse numbers reflects the fact that in Sanskrit, the phonetic unit of writing is the syllable rather than the individual vowel or consonant. Verses with the same number of syllables tend to take up about the same amount of linear space to write out and thus tend to fall into periodic-looking patterns when written.) Other than that, the text is not formatted in any way but rather reproduced on the page as a stream of syllables, just as it would be recited orally.

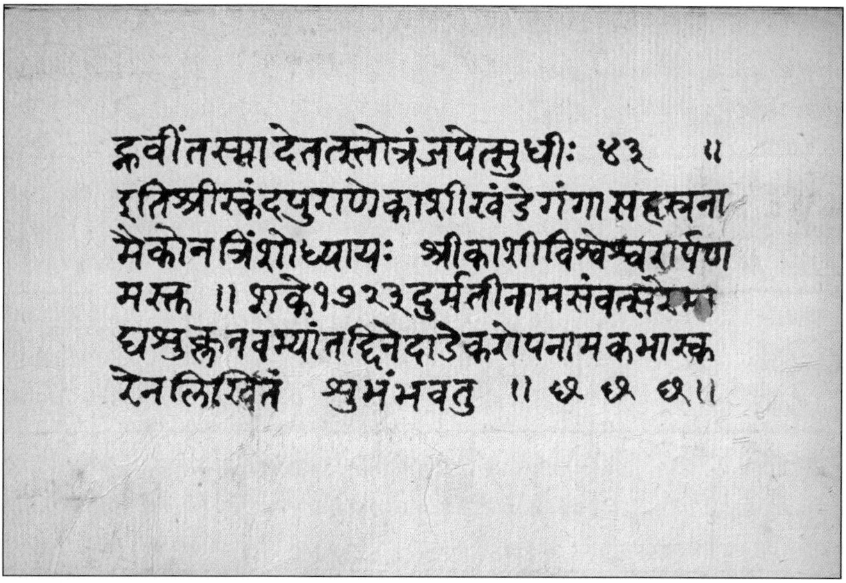

क्तवींतस्मादेतत्स्तोत्रंजपेत्सुधीः ४३ ॥
इतिश्रीस्कंदपुराणेकाशीखंडेगंगासहस्रना
मैकोनत्रिंशोध्यायः श्रीकाशीविश्वेश्वरार्पण
मस्तु ॥ शके १७२३ दुर्मतीनामसंवत्सरेमा
घशुक्लनवम्यांतद्दिनेदाडेकरोपनामकभास्क
रेनलिखितं शुभंभवतु ॥ छ छ छ ॥

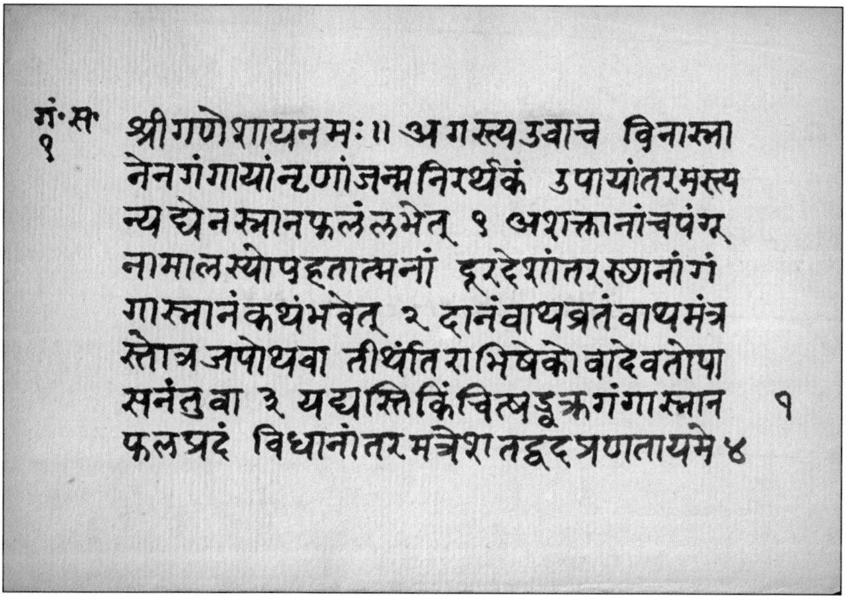

गं·स· १ श्रीगणेशायनमः ॥ अगस्त्यउवाच विनास्ना
नेनगंगायांनृणांजन्मनिरर्थकं उपायांतरमस्त्य
न्यद्येनस्नानफलंलभेत् १ अशक्तानांचपंगू
नामालस्योपहतात्मनां दूरदेशांतरस्थानांगं
गास्नानंकथंभवेत् २ दानंवाथव्रतंवाथमंत्र
स्तोत्रजपोथवा तीर्थांतराभिषेकोवादेवतोपा
सनंतुवा ३ यद्यस्तिकिंचित्षड्वक्त्रगंगास्नान १
फलप्रदं विधानांतरमत्रेशातद्वदप्रणतायमे ४

Figure 7.1. First page (fol. 1v) of the Purāṇa text *Gaṅgā-sahasra*. Philadelphia, University of Pennsylvania Libraries, Penniman-Gribell Collection, no. 2332.

Figure 7.2. Final page (fol. 28r) of the *Gaṅgā-sahasra*. Philadelphia, University of Pennsylvania Libraries, Penniman-Gribell Collection, no. 2332.

The end of the manuscript, as al-Bīrūnī remarks, is where the identification of its content appears. The end of the abovementioned manuscript is shown in Figure 7.2. After the final verse number (43), the text concludes: "Thus the twenty-ninth chapter called *Gaṅgā-sahasra* in the *Kāśī-khaṇḍa* in the revered *Skanda-purāṇa*. May it be an offering to Kāśī-Viśveśvara [Śiva]. In the year Śaka 1723 [1801 CE], called Durmatī [in the sixty-year Jupiter cycle], on the ninth [lunar] day of the waxing fortnight of [the month] Māgha, [it was] written [copied] by Bhāskara surnamed Dāṃḍekara. May it be auspicious." Here the text of the work is primarily a spoken composition, with what might be called a lapse into conscious literacy at the end.

The purely oral format is not ideally suited to scientific texts, especially to their expository apparatus of commentary, worked examples, tables and figures, and, in particular, numerical values. Number words are notoriously difficult to fit into a metrical verse structure since they are so rigid in their meaning and format. Indian scientists around the first few centuries of this era circumvented this problem in writing scientific verse by means of the so-called *bhūta-saṅkhyā*, or "object number" system,[10] described by al-Bīrūnī as follows: "For each number they have appropriated quite a great quantity of words. Hence, if one word does not suit the metre, you may easily exchange it for a synonym which suits. [Pseudo]-Brahmagupta says: 'If you want to write *one*, express it by everything which is unique, as *the earth*, *the moon*; *two* by everything which is double, as, *e.g., black and white*; *three* by everything which is threefold; the *nought* by *heaven*, the *twelve* by the names of the sun.'"[11] In manuscripts, such "concrete numbers" were generally accompanied by their written numeral equivalents, as in the examples seen in Figure 7.3, which shows a page from an 1820 copy of an astronomical treatise composed a few decades earlier. Near the start of the second line is the compound *kha-aṅga-agni*, "void-limb-fire," immediately followed by the numerals "360." "Void" signifies zero; "limb," six (from the conventional six limbs or supporting disciplines of the sacred Veda); and "fire," three (from the three fires used in worship rituals). Concrete numbers and original Sanskrit number words could be mixed in such compounds, as in the first word of line 3, *kha-aṣṭa-bhū*, "void-eight-earth," "180."

The design of this concrete-number system, a postliterate development in Sanskrit verse, seems to have been inspired specifically by the written

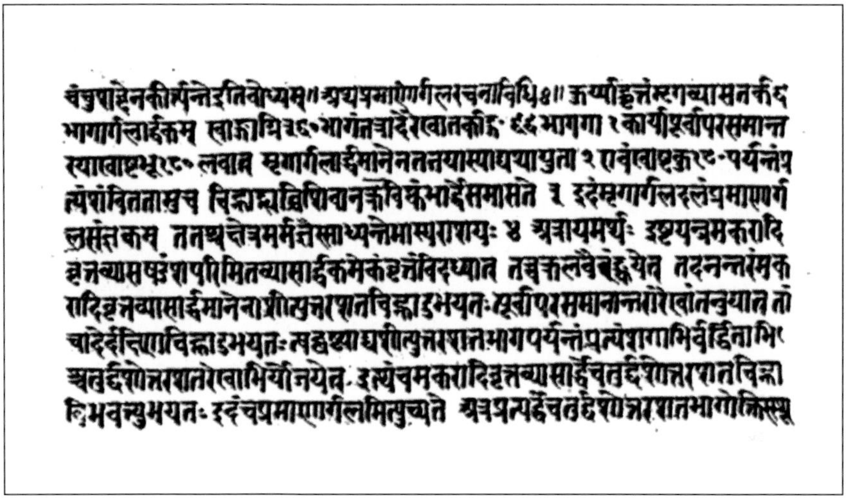

***Figure* 7.3.** Versified number words and numerals in an astronomical manuscript. Varanasi, Sampurnanand Sanskrit University, no. 35245.

form of numerals in decimal place-value notation: "zero-six-three" rather than the traditional verbal presentation "three hundred and sixty." It is not clear why the digit stream of a concrete number was ordered from the least significant to the most significant digit rather than the other way around. A decimal place-value numeral would naturally have been written in left-to-right Indic scripts with its most significant digit first. Whatever its origin, this least-to-most-significant ordering seems to have been fixed in the concrete-number notation before its earliest recorded use. And, of course, an absolutely rigid ordering convention would be necessary for the notation to be useful, as there is no way to tell from the strings of verbally encoded digits themselves whether the first or the last is supposed to represent the unit's place.

As seen in the examples in Figure 7.3, numerals accompanying concrete-number words in manuscripts were simply inserted into the text stream, not graphically distinguished in any way from verse numbers. Moreover, the verse numbers themselves might temporarily disappear from the text stream (as in the page shown in the figure, where the diagonal pattern of verse numbers 1-2-3-4 is not continued by a 5) because the sequence of numbered verses in the base text was interrupted by a patch of commentary in prose. The written text is thus a sometimes confusing mix of oral and visual features, maintaining a linear flow.

The strong strain of orality persisting in this manuscript tradition may have exciting implications for the study of pedagogical methods in Sanskrit science. The late David Pingree maintained that the style of certain astronomical commentaries strongly suggests that they were essentially verbatim transcriptions of oral instruction: class lecture notes, as it were.[12] We are used to thinking of a commentary as a writer's engagement with a written text; what we may have in addition within the Sanskrit scientific tradition is the alternative concept of a commentary as a written record of a spoken lecture.

We cannot infer too much from this possibility, however, since we also frequently see scientific manuscripts that are clearly not continuous verbal productions but contain more temporal layers, as illustrated in Figure 7.4. This manuscript of a sixteenth-century astronomical handbook shows a purely visual feature, a table placed in a box display to one side of the text stream. As is usual in Sanskrit scientific manuscripts but less so in Western ones, the table is not referenced or described in the text itself: it is literally a silent accompaniment to the words of the book.

We can also see evidence of a scribe's or a reader's choices in the addition of numerals above the corresponding number words occurring in the verses as originally copied. In line 2, there is a small "16" inserted above the word *nṛpa*, or "ruler," a conventional concrete number for sixteen. In the line below, "15" is similarly written in above the concrete number *śara-ku*, "arrow-earth" (where "arrow" stands for the metaphorical five arrows of the love-god Kāma).

The insertion of numerals as an artifact of the scribal process is also attested by careless placement of them where they are not appropriate. For example, a scribe will sometimes encounter the word "earth" or "moon" and write the numeral "1" following it, even when the word in the text as originally composed is meant solely in its physical sense and not as a number.

This blurring of the roles of word and numeral can have repercussions in scribal practice where it is not completely clear which numbers accompany

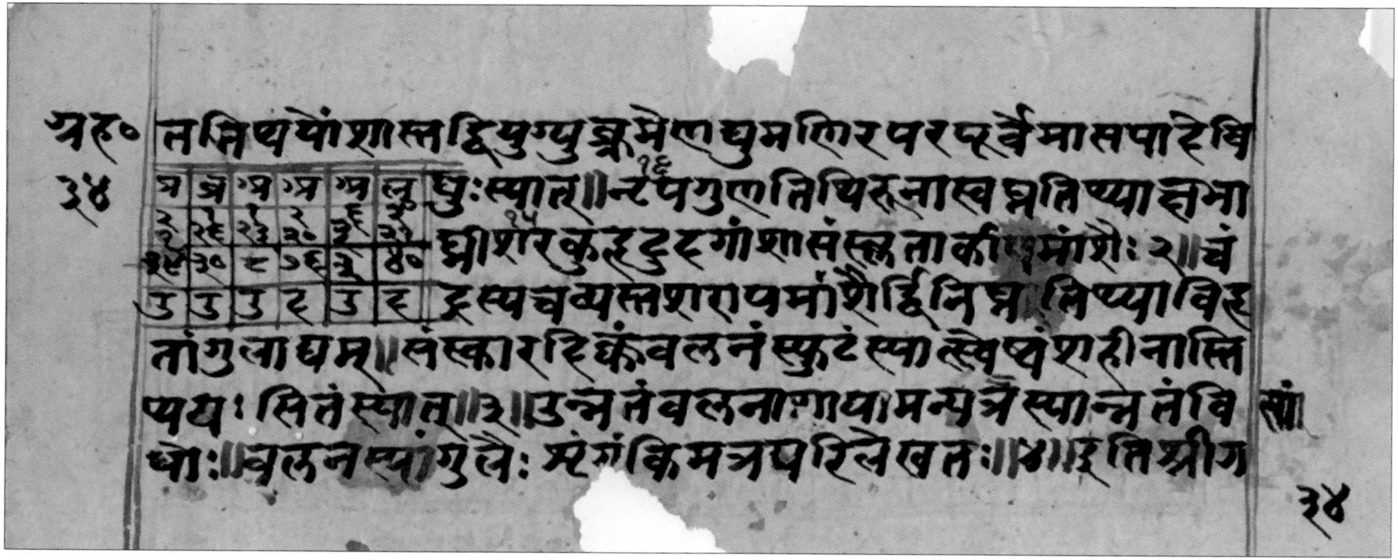

Figure 7.4. Numerical tables and decoded numerals in an astronomical handbook. Commentary on Graha-lāghava of Gaṇeśa. Private collection, fol. 34r.

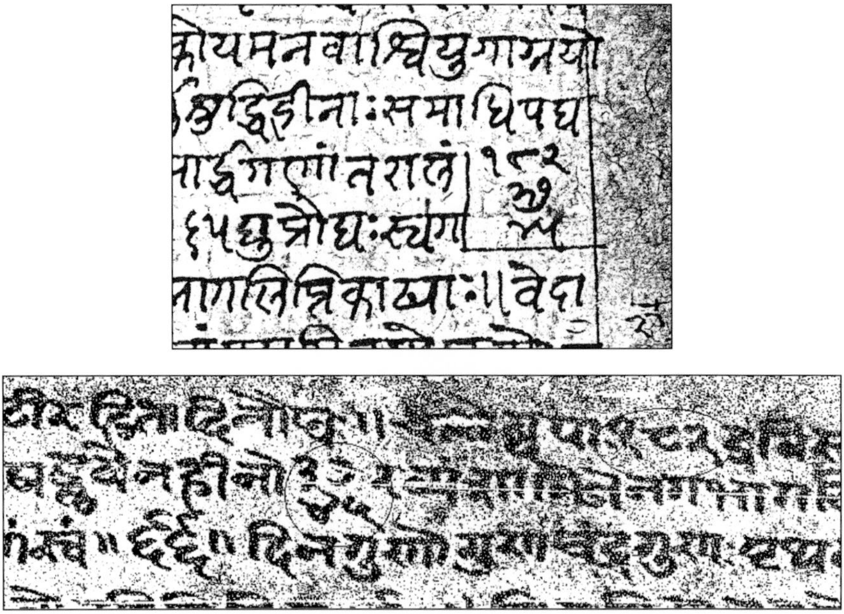

Figure 7.5. Displayed numbers correctly and incorrectly copied in two manuscripts of the same work (top: Varanasi, Sampurnanand Sanskrit University, no. 35566, fol. 39v; bottom: Pune, Bhandarkar Oriental Research Institute, no. 860 of 1887–91, fol. 5v).

specific words in a text and which are stand-alone displays. Figure 7.5 shows partial images of leaves from two manuscripts of the same text, which display the same group of numbers: 182, 37, 45. But in the second manuscript, the scribe has mistakenly perceived these numbers in his source as belonging with different words on different lines of the text rather than grouped separately in a display. So he has copied them in different places, on the first and second lines of the excerpt.[13] This sort of misreading is not uncommon and can render the role of the copied numerals incomprehensible in one or two iterations of miscopying. The scribal "carelessness" that al-Bīrūnī grumbled about seems in this case to signify a lesser importance attributed to these appended numerals; they are carried along with the text stream but are not really seen as part of its structure, and their inclusion and placement are not crucial to the correctness of the words of the text.

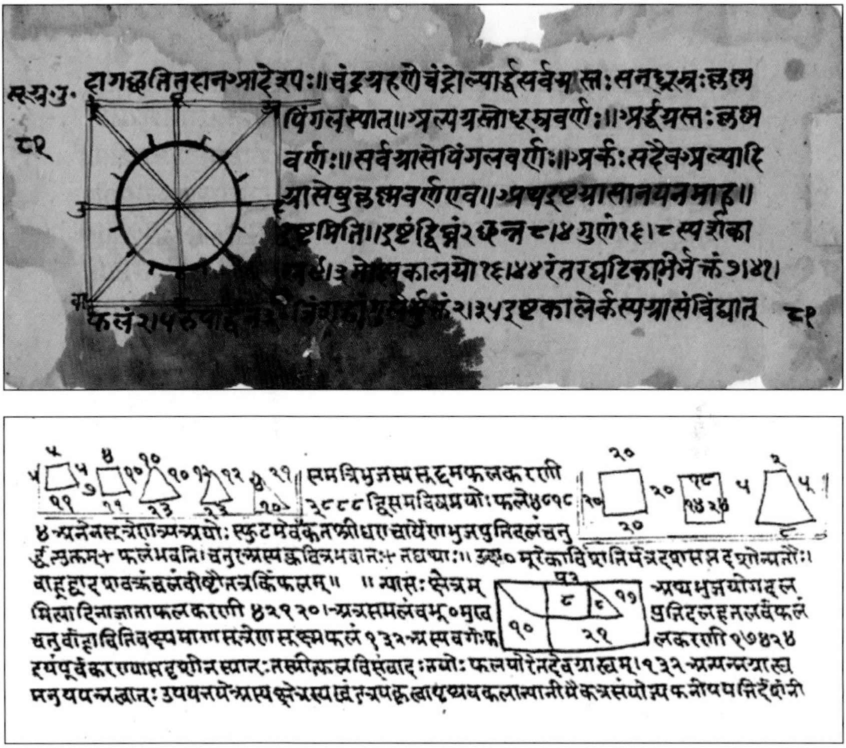

Figure 7.6. Diagrams in an astronomical manuscript (fol. 81v in the ms. of Figure 7.4) and a mathematical one (Varanasi, Sampurnanand Sanskrit University, no. 104595, fol. 58v).

The remaining major type of nonverbal feature in a scientific manuscript is the technical diagram. In the indigenous Sanskrit scientific tradition, diagrams tend to be few in number, and they do not interact with the text content in the way that, say, the geometric figures of Euclid do. As illustrated in the examples in Figure 7.6, they are occasional visual reinforcements for verbal explanations and rules, and they are generally roughly schematic rather than precisely traced. The figure in the first manuscript represents the beginnings of an eclipse diagram (which the scribe apparently never completed), while those in the second are modeling various geometry formulas. As with displayed numbers, a chunk of the page's area is set aside

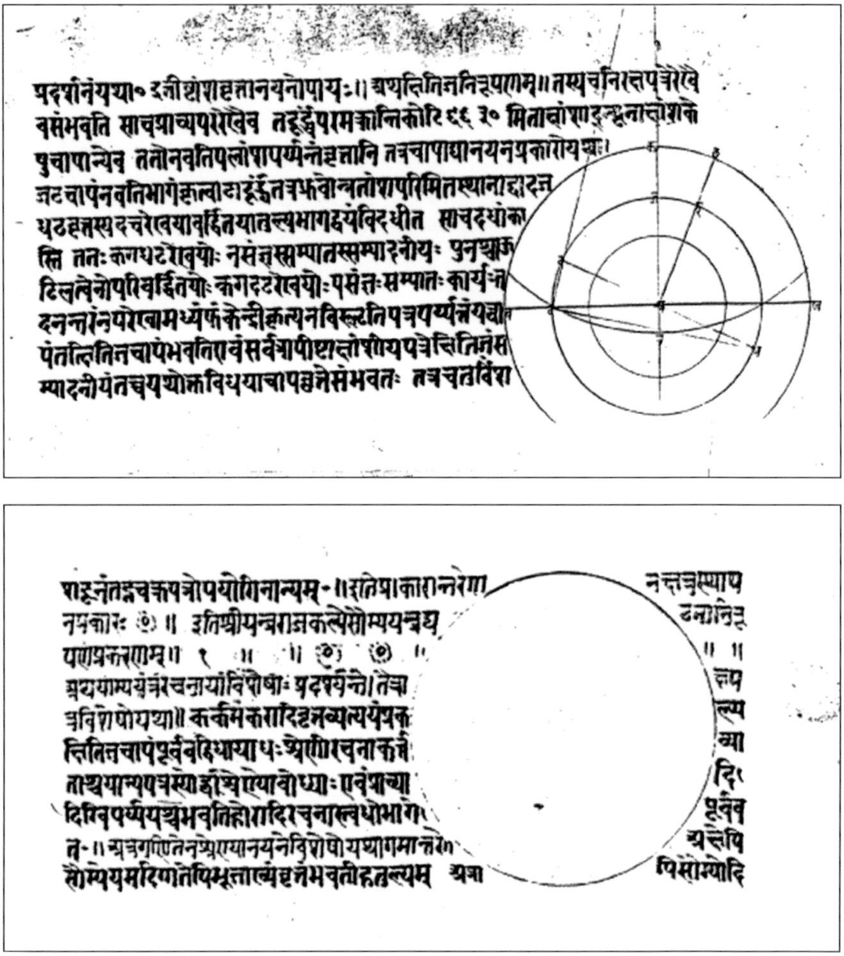

Figure 7.7. Diagrams in a late work (the ms. of Figure 7.3), showing Islamic influence (top: fol. 4r, bottom: fol. 26r).

for diagrams in apparently random segments, and the text stream with its own incidental numbers flows beside and around those spaces. Again as with numbers, the occasional unfinished or misrepresented form of a figure seems to be considered a tolerable omission.

The same casualness about the role of figures is still observed, although to a lesser extent (Figure 7.7), in a manuscript of a late work, the same previously shown in Figure 7.3. In this late eighteenth-century text describing and explaining the Islamic astrolabe, although most of the style and layout is typical of traditional Sanskrit manuscripts, elements of Islamic influence can be seen in the drawings. They are carefully formed with compass and straight-edge, not casually sketched freehand. And the key points in the first figure are labeled with letters of the alphabet, a practice that Arabic and Persian authors picked up from Greek geometry texts but that Sanskrit scholars did not use until they encountered it in Islamic texts. Note that the diagram in the second leaf is blank; Indian scribes still frequently failed to fill in spaces left for features like tables and figures.

The tradition of the Indian scientific manuscript was by this time within a century and a half of its end. As colonial administrators strove to foster modern Western education and technology, and as Indian scientists sought recognition in European and eventually global institutions, the Sanskrit scientific manuscript fell out of favor as a way to preserve and transmit technical learning. While certain works (primarily sacred texts) were and in some cases still are printed in traditional format on oblong pages, the Western-style book soon superseded manuscripts in all other genres, including the exact sciences. Fittingly, after their long cooperation, memorized oral learning and written manuscripts in the sciences faded from the scene together.

Notes

1 *Cāṇakya-nīti-śāstra*, 16.20.

2 Edward Sachau, *Alberuni's India,* 2 vols. (New Delhi: Munshiram Manoharlal, repr. 1992), 19.

3 Ibid., 125–26.

4 Ibid., 136–37.

5 William M. Calder and Stephan Heilen, "David E. Pingree: An Unpublished Autobiography," *Greek, Roman and Byzantine Studies* 47 (2007), 522; B. N. Goswamy, *The Word Is Sacred, Sacred Is the Word: The Indian Manuscript Tradition* (London: Philip Wilson, 2008). Thirty million was the total estimated by David Pingree; a more conservative figure of approximately five million manuscripts in India alone is suggested in Goswamy, *Sacred*, 17.

6 Sachau, *Alberuni's India* (note 2), 171.

7 Ibid., 18.

8 Ibid., 173.

9 Ibid., 182.

10 K. V. Sarma, "Word and Alphabetic Numeral Systems in India," in *The Concept of Śūnya,* ed. A. K. Bag and S. R. Sarma, 38–41 (Delhi: National Science Academy, 2003).

11 Sachau, *Alberuni's India* (note 2), 177.

12 David Pingree, "Special 'Schools,'" in *Enciclopedia Italiana: Storia della scienza. La scienza indiana* (Rome: Enciclopedia Italiana Treccani, 2004).

13 Kim Plofker, "Sanskrit Mathematical Verse," in *The Oxford Handbook of the History of Mathematics,* ed. Jacqueline Stedall et al., 533 (Oxford: Oxford University Press, 2008). The same example is more briefly described here.

CHAPTER 8

Abbreviations in Medieval Astronomical and Astrological Manuscripts Written in Arabic Script

Sergei Tourkin

Abbreviations were a common practice in handwritten texts transcribed in the Arabic script during the manuscript age.[1] The basic principles of how words and phrases were abbreviated in these texts are known to us.[2] However, because each subject used a particular terminology of its own, specific abbreviations were employed to reflect a given contextual setting, and therefore dissimilar abbreviations were used for different subjects. On the other hand, some abbreviations, which look exactly the same, could be used to shorten different words, depending on the subject of a text.

This essay discusses the abbreviations used in Persian and Arabic manuscripts containing mainly, but not exclusively, works on astronomy and astrology.[3] Abbreviations particularly abound in two types of astronomical-astrological compositions: almanacs (year calendars arranged on a monthly basis) and personal horoscopes (cast for a person's time of birth). Both types of composition are still relatively little known; they chiefly contain results of astronomical observations and calculations together with astrological predictions based on them. Often the very page layout of the text in such works—especially when the data were arranged in tabular form—necessitated the use of abbreviations because the cells in a table grid, used for certain astronomical parameters or astrological properties, were limited by the height of the rows and the width of the columns.

This study is based on four medieval astronomical-astrological works in Persian that deal with the compilation of almanacs. These works are (in chronological order)

1. An anonymous Persian translation of *Al-madḫal ilá ᶜilm aḥkām al-nujūm* by Abū-Naṣr al-Ḥasan b. ᶜAlī Qumī (written in 365 AH [975–76 CE]; hereafter referred to as Qumī, with page references to the edition of the text by Jalīl Aḫavān-Zanjānī.)[4]
2. *Rawżat al-munajjimīn* or *Rawżat al-munajjim* by Šahmardān b. Abī-al-Ḫayr Rāzī (written in 466 AH [1073–74 CE]; hereafter referred to as Rāzī, with page references to the edition of the text by Jalīl Aḫavān-Zanjānī)[5]
3. *Muḫtaṣar dar maᶜrifat-i taqwīm* (also known under the title *Sī faṣl dar [maᶜrifat-i] taqwīm*) by Naṣīr-al-Dīn Ṭūsī (d. 672 AH [1274 CE]; hereafter referred to as Ṭūsī, with page references to the facsimile edition of the text in the *Safīna-yi Tabrīz*, which is dated 12 Rabīᶜ I 723 AH [20 March 1323 CE] and contains one of the earliest known copies of Ṭūsī's work)[6]
4. *Muḫtaṣar dar maᶜrifat-i taqwīm wa usṭurlāb* by Ġiyāṯ-al-Dīn Maḥmūd Kāšānī (d. 832 AH [1429 CE] or 839 AH [1436 CE]; hereafter referred to as Kāšānī, with folio references to London, British Library Add. Ms. 27261 [fols. 365v–369r])

When encountering abbreviations in astronomical and astrological texts, the reader's first inclination in deciphering them may well be to apply one or both of the widely used systems for rendering astronomical coordinates and other numerical values: the alphanumerical notation called *abjad* and the system of representing compound numbers in writing named *jummal*. However, for the purpose of abbreviating words and phrases as such, the *abjad-jummal* notation turns out to be applicable in only a small number of cases, such as when the abbreviated items make a logical linear sequence and can therefore be numbered or counted from a conventional starting point rather than having their names abbreviated. Thus, in the majority of instances of abbreviations (excluding the aforementioned astronomical coordinates and other numeric values), the *abjad-jummal* notation offers no help in understanding which words or phrases have been abbreviated. Even in the case of the names of the planets, which could be arranged in a "linear" sequence, either ascending or descending, depending upon their distances from Earth (which, according to the geocentric system, was believed to be located at the center of the universe), the abbreviations used for the names of the planets have nothing to do with the *abjad-jummal* notation, as will be shown below.

Let us start with the twelve zodiacal signs. Their names were abbreviated using the *abjad-jummal* notation as follows: Aries equals the sign for zero, Taurus is *alif* (one), Gemini is *bā'* (two), Cancer is *jīm* (three), Leo is *dāl* (four), and so on, up to Pisces, which is *yā'-alif* (eleven).[7] The adjectives *šimālī* (northern) and *janūbī* (southern), when applied to the zodiacal signs, were abbreviated by their first letters as *šīn* and *jīm* accordingly.[8]

As in the case of the zodiacal signs, the *abjad-jummal* notation was likewise used for abbreviating the days of the week: Sunday is *alif* (one), Monday is *bā'* (two), Tuesday is *jīm* (three), Wednesday is *dāl* (four), Thursday is *hā'* (five), and Friday is *wāw* (six). Saturday could be abbreviated either by the letter *zāy* (seven) or by the special sign for zero.[9] But this is where the use of the *abjad-jummal* notation for abbreviations in this field comes to an end.

When we come to the names of the seven planets,[10] we find that their names were abbreviated in a quite different way. The usual names of the seven planets in Arabic (and also in Persian, except that in Persian the Arabic definite article *al-* was often omitted) are (in ascending order from Earth) *al-qamar* (Moon), *ʿuṭārid* (Mercury), *al-zuhrah* (Venus), *al-šams* (Sun), *al-mirrīḫ* (Mars), *al-muštarī* (Jupiter), and *zuḥal* (Saturn). Simple observation reveals the fact that it is not possible to abbreviate the names of the planets by their first letters since the names of two planets start with the letter *zāy* (*al-zuhrah* and *zuḥal*) and the names of two other planets begin with the letter *mīm* (*al-mirrīḫ* and *al-muštarī*). Because of this, the names of the planets were abbreviated using their last letters:[11] *al-qamar* (Moon) is *rā'*, *ʿuṭārid* (Mercury) is *dāl*, *al-zuhrah* (Venus) is *hā'*, *al-šams* (Sun) is *sīn*, *al-mirrīḫ* (Mars) is *ḫā'*, *al-muštarī* (Jupiter) is *yā'*, and *zuḥal* (Saturn) is *lām*.[12] Such a solution turns out to be both easy and effective.

The terms *ṣāʿid* (ascending) and *hābiṭ* (descending), when applied to heavenly bodies, were abbreviated using their first letters as *ṣād* and *hāʾ*, respectively.[13] However, the terms *zāyid* (increasing) and *nāqiṣ* (decreasing) were abbreviated using their last two letters as *yāʾ-dāl* and *qāf-ṣād*, respectively (the short vowel *i* in both words is not written in the Arabic script). For the abbreviations of these two pairs of terms, Ṭūṣī explains the abbreviations in clear words, as *ḥarf-i awwal* (the first letter) and *dū ḥarf-i āḫar* (the last two letters).[14]

The special cases of the locations of planets on the ecliptic with regard to one another are called "aspects" (*naẓar*, pl.: *anẓār*; lit.: "look") in astrology and include the conjunction (zero-degree difference between a pair of planets), the sextile (60-degree difference), the quartile or quadrature (90-degree difference), the trine (120-degree difference), and the opposition (180-degree difference). The names of the different aspects of the planets were abbreviated using their last letters only: conjunction (*muqāranah* or *qirān*) was abbreviated as *nūn*, conjunction of the Sun and the Moon (*ijtimāʿ*) as *ʿayn*, sextile (*tasdīs*) as *sīn*, quartile (*tarbīʿ*) as *ʿayn*, trine (*taṯlīṯ*) as *ṯāʾ*, and opposition (*muqābalah* or *istiqbāl*) as *lām*.[15] Here, as in the case of the names of the planets, three terms start with the same letter *tāʾ* (*tasdīs*, *tarbīʿ*, and *taṯlīṯ*), while another two terms begin with the letter *mīm* (*muqāranah* and *muqābalah*), which excludes the possibility of using the first letters for abbreviating these terms. Moreover, in all these words, neither the initial letter *tāʾ* nor the initial letter *mīm* belongs to the semantic Arabic root; their presence at the beginning of all the above-mentioned words is stipulated by Arabic morphological rules. The same can be said of the words *muqāranah* and *muqābalah*, whose endings are the traditional endings of the feminine gender. Their presence is stipulated by the grammar, and they do not carry any special semantic load; exactly the same endings are found at the end of all feminine parts of speech in Arabic, which makes their use in abbreviations somewhat problematic if not impossible. For all these reasons, the accepted abbreviations for the words *muqāranah* and *muqābalah* (*nūn* and *lām*) are not, strictly speaking, their last letters but the last radical consonants of the roots from which these two words originate.

Conjunctions of the planets with the two nodes (two extreme points) of the lunar orbit were referred to as *mujāsadah*, which was abbreviated as *mīm-jīm-alif.*[16] The names of these two nodes, or extreme points, in Arabic are *al-ra's* and *al-ḏanab* (the head and the tail), and both of them could either be abbreviated using their last two letters as *alif-rā'* and *nūn-bā'* (the second *a* in *al-ḏanab* is not written in Arabic)[17] or spelled in full to avoid confusion.

Ṭūṣī writes that the terms *muqīm* (stationary) and *mustaqīm* (progressive), both applied to the types of motions of the planets, are written in full (the two terms begin and also end with the same letter *mīm*), while the term *rāji*ᶜ (retrogressive) is abbreviated using its last two letters, *jīm-*ᶜ*ayn* (the *i* is not written in Arabic).[18] However, according to Kāšānī, both *mustaqīm* (progressive) and *rāji*ᶜ (retrogressive) are to be shortened by their last letter only (*mīm* and ᶜ*ayn*, respectively), and he does not make any mention of the term *muqīm* (stationary).[19]

Other astrological terms (explanations of their meanings are beyond the scope of this study and can be found in manuals on astrology) were abbreviated using their last letters as follows: *šaraf* (exaltation) as *fā'*; *hubūṭ* (dejection) as *ṭā'* (see above for the use of the first letter in abbreviating the word *hābiṭ*, which derives from the same Arabic root); *iḥtirāq* (combustion) as *qāf*; *intikāṯ* (reversal from a planetary aspect) as *ṯā'*; *tadwīr* (epicycle) as *rā'*; *awj* (apogee) as *jīm*; *šarq* and *mašriq* (east) as *qāf*; *ġarb* and *maġrib* (west) as *bā'*; *taḥwīl* (shift or transfer from one zodiacal sign to another) as *yā'-lām*; and *tanāẓur* (symmetry in the positions of two planets) as *ẓā'-rā'*.[20]

While the letter *jīm* is said to be, and indeed often is, the abbreviation for the word *awj* (apogee), none of the sources that I have consulted for this study mentions any abbreviation for the word *ḥaḍīḍ* (perigee). From my own experience of looking at manuscripts with almanacs and horoscopes, I can attest that the word *ḥaḍīḍ* (perigee), when not written in full, which is relatively frequently, could be abbreviated either using its last letter *ḍād* or using the combination of its first letter *ḥā'* and the letter *ḍād*. However, since the word *ḥaḍīḍ* has two letters *ḍād*, the question arises whether the letter *ḍād* in this two-letter abbreviation *ḥā'-ḍād* is the second letter of the word *ḥaḍīḍ* or its last letter. I would prefer to view it as the last letter of the word so that in this case we have an enveloping or circumscribing type of abbreviation when a word is abbreviated using its first and last letters. In support of my view, I have also come across many cases where the word *šaraf* (exaltation) is abbreviated using its first and last letters as *šīn-fā'* and not its last letter *fā'* only as is recommended in the sources. A similar, if not analogous way of abbreviating words and phrases can be seen in texts on other subjects as well.[21]

The sectors (*niṭāq*, pl.: *niṭāqāt*) of the planetary orbs were abbreviated using several letters. The last letter (*qāf*) in the word *niṭāq* (sector) was first followed by the *abjad* sign for the number of the sector on an orb (from one to four, represented by the letters *alif* to *dāl*) and then by the letter *jīm* if the orb in question was the apogee (*awj*), or by the letter *rā'* if the orb was the epicycle (*tadwīr*). An important point here is that the combination of the letter *qāf* and a letter from *alif* to *dāl* (for the number of this sector) was written separately from either the letter *jīm* or the letter *rā'*,[22] both of which were written in their isolated form.

The word *aysar* (left or left hand, when applied to the types of aspects of the planets) could be abbreviated using the last letter *rā'*, whereas the word *ayman* (right or right hand, also when applied to the types of aspects of the planets) was always written in full. This must have been so decided because the last letter (*nūn*) in the word *ayman* (right) was already used to designate the conjunction of a pair of planets (*qirān* or *muqārana*), despite the fact that the last letter (*rā'*) in the word *aysar* (left) was likewise already used, and even more widely, as the abbreviation for the words "moon" (*al-qamar*), "epicycle" (*tadwīr*), and "daytime" (*nahār*; see Table 8.2). In most cases, the context seems to make it quite clear which word has been abbreviated by a letter despite its being used to represent several terms.

Besides the "pure" astronomical and astrological terms, some of those related to geography also had their abbreviations. As already mentioned above, the words "east" (*šarq* or *mašriq*) and "west" (*ġarb* or *maġrib*) were abbreviated using their last letters, *qāf* and *bā'*.[23] The words *balad* (city or locality), *mawḍiʿ* (place), *jabal* (mountain), and *qaryah* (village) were all abbreviated using their last letters as *dāl*, *ʿayn*, *lām*, and *hā'*, respectively.[24]

Certain terms relating to chronology could be abbreviated as well. The words *yawm* (day) and *laylah* (night) were abbreviated using their last letters (for *laylah*, see the remark above regarding the words *muqāranah* and *muqābalah*) as *mīm* and *lām*.[25] Similarly, the words *nahār* (daytime) and *layl* (nighttime) were abbreviated using their last letters as *rā'* and *lām*.[26]

Each month of the Hijra year also had its own abbreviations. In fact, ten out of the twelve months had two or three abbreviations each. These abbreviations have been analyzed and published by Adam Gacek,[27] and the table below, compiled on the basis of his research, is given here for the purpose of reference and illustration.

Table 8.1

Month	Abbreviations	Month	Abbreviations
1. Muḥarram	*mīm*	7. Rajab	*bā'*; *rā'*
2. Ṣafar	*ṣād*	8. Šaᶜbān	*šīn*; *šīn-ᶜayn*
3. Rabīᶜ I	*rā'-1*; *ᶜayn-1*; *ᶜayn-lām*	9. Ramaḍān	*nūn*; *mīm-ḍāḍ*
4. Rabīᶜ II	*rā'*; *rā'-2*; *ᶜayn-2*	10. Šawwāl	*šīn*; *lām*
5. Jumādá I	*jīm*; *jīm-1*; *jīm-alif*	11. Ḏū-al-Qaᶜdah	*ḏāl-1* (or: *ḏāl-alif*); *qāf-ᶜayn*
6. Jumādá II	*jīm*; *jīm-2*	12. Ḏū-al-Ḥijjah	*ḏāl*

As demonstrated by many of the above examples, quite often it was the last letter in a word (or the last radical consonant of the Arabic root) rather than the first letter that was used to construct abbreviations in manuscripts written in Arabic script in the Middle Ages. A continuation of this method of abbreviating can still, albeit rarely, be found in modern times too: in many Persian catalogues of manuscripts printed in the twentieth and twenty-first centuries, for example, the common abbreviation for the word *barg* (folio in Persian) is its last letter *gāf*.

However, nowadays the general practice of using the last letters for abbreviating Arabic-script words appears to have greatly decreased, if not stopped completely. Thus, for example, the overwhelming majority of modern abbreviations accepted in the Arabic language, which are provided in Māmaqānī's study,[28] are the first letters of the abbreviated words. Moreover, in many modern editions of medieval Arabic dictionaries the original order of Arabic roots undergoes drastic changes. Thus, while in the original works the roots are primarily organized on the basis of an alphabetical arrangement of their last radicals (root characters), and only then taking into account their middle and first radicals, modern editions rearrange the entire sequence of the roots to conform to the direct alphabetical order of the radicals, starting from the first.[29]

This study reflects only the initial stage of my research into medieval Arabographic manuscripts (i.e., manuscripts written in the Arabic script but not necessarily in the Arabic language). For the time being, I have restricted myself to the subjects of astronomy and astrology, as reflected in the title of the essay.

Table 8.2 Summary Table of Abbreviations Mentioned in This Study

Arabic letter(s) of abbreviations	Abbreviated words or phrases	Meaning/ Translation
sign for zero		1. Aries 2. Saturday (optional)
alif	1. *abjad* value = 1 2. *abjad* value = 1	1. Taurus 2. Sunday
alif-rā'	*al-ra's*	"The Head" (ascending lunar node)

Arabic letter(s) of abbreviations	Abbreviated words or phrases	Meaning/ Translation
bā'	1. *abjad* value = 2 2. *abjad* value = 2 3. *ġarb* or *maġrib*	1. Gemini 2. Monday 3. West
ṯā'	*taṯlīṯ* *intikāṯ*	1. trine 2. reversal from a planetary aspect
jīm	1. *abjad* value = 3 2. *abjad* value = 3 3. *awj* 4. *janūbī*	1. Cancer 2. Tuesday 3. apogee 4. southern
ḥā'	1. *abjad* value = 8	Sagittarius
ḥā'-ḍāḍ	*ḥaḍīḍ*	perigee
ḫā'	*al-mirrīḫ*	Mars
dāl	1. *abjad* value = 4 2. *abjad* value = 4 3. *ʿuṭārid* 4. *balad*	1. Leo 2. Wednesday 3. Mercury 4. city or locality
rā'	1. *al-qamar* 2. *tadvīr* 3. *nahār* 4. *aysar*	1. Moon 2. epicycle 3. daytime 4. left
zāy	1. *abjad* value = 7 2. *abjad* value = 7	1. Saturday 2. Scorpio

continued

Table 8.2 (continued)

Arabic letter(s) of abbreviations	Abbreviated words or phrases	Meaning/ Translation
sīn	1. *al-šams* 2. *tasdīs*	1. Sun 2. sextile
šīn	*šimālī*	northern
šīn-fā'	*šaraf*	exaltation
ṣād	*ṣā*ᶜ*id*	ascending (heavenly body)
ḍād	*ḥaḍīḍ*	perigee
ṭā'	1. *abjad* value = 9 2. *hubūṭ*	1. Capricorn 2. dejection
ẓā'-ra'	*tanāẓur*	symmetry (in the positions of a pair of planets)
ᶜayn	1. *tarbī*ᶜ 2. *rājiᶜ* 3. *mawḍiᶜ* 4. *qāṭiᶜ* (not mentioned in the article)	1. quartile (quadrature) 2. retrogressive (motion) 3. place 4. cutter [of life], anareta or interfector
fā'	*šaraf*	exaltation
qāf	1. *šarq* or *mašriq* 2. *iḥtirāq* 3. *niṭāq*	1. east 2. combustion 3. sector (of an orb)

Arabic letter(s) of abbreviations	Abbreviated words or phrases	Meaning/ Translation
qāf-ṣād	*nāqiṣ*	decreasing
gāf	*barg* (Persian)	folio
lām	1. *zuḥal* 2. *laylah* 3. *layl* 4. *jabal*	1. Saturn 2. night 3. night-time 4. mountain
mīm	1. *mustaqīm* 2. *yawm*	1. progressive (motion) 2. day
mīm-jīm-alif	*mujāsadah*	conjunction of a planet with a lunar node
nūn	*qirān* or *muqāranah*	conjunction of planets
nūn-bā'	*al-ḏanab*	"The Tail" (descending lunar node)
hā'	1. *abjad* value = 5 2. *abjad* value = 5 3. *al-zuhrah* 4. *hābiṭ* 5. *qaryah*	1. Virgo 2. Thursday 3. Venus 4. descending (heavenly body) 5. village
wāw	1. *abjad* value = 6 2. *abjad* value = 6	1. Libra 2. Friday

continued

Table 8.2 (continued)

Arabic letter(s) of abbreviations	Abbreviated words or phrases	Meaning/ Translation
yā'	1. *abjad* value = 10 2. *al-muštarī*	1. Aquarius 2. Jupiter
yā'-alif	*jummal* value = 11	Pisces
yā'-dāl	*zāyid*	increasing
yā'-lām	*taḥwīl*	shift, transfer (of a planet to the next zodiacal sign)

Notes

I am very grateful to Stephen Millier, David Nancekivell, and Adam Gacek (all from McGill University, Montreal) for their valuable suggestions regarding both the English language and the content of this essay. None of them is responsible for any mistake that this essay may contain.

1 In this essay, the word *abbreviation* is used in its wider and generic sense of shortening a word or a phrase, without differentiating between specific types of abbreviations, such as acronyms, initialisms, sigla, contractions, suspensions, and others.

2 Adam Gacek, "Abbreviations," in *Encyclopedia of Arabic Language and Linguistics*, ed. Kees Versteegh, vol. 1, 1–5 (Leiden: E. J. Brill, 2006). Gacek's article is, to date, the most comprehensive overview of abbreviations in texts written in the Arabic script.

3 A brief list of abbreviations for astronomical and astrological terms, some of which are discussed in this essay, is given in Muḥammad-Riḍā Māmaqānī, *Muʿjam al-rumūz wa al-išārāt*, 2nd ed. (Beirut: Dār al-Muʾarriḫ al-ʿArabī, 1992), 67–69. However, there is no explanation or analysis for the abbreviations listed in this publication.

4 *Tarjuma-yi al-Madḫal ilá ʿilm aḥkām al-nujūm (ta'līf ba sāl-i 365 H.Q.) Abū-Naṣr Ḥasan b. ʿAlī Qumī az mutarjim-ī nāšināḫta*, ed. Jalīl Aḫavān-Zanjānī (Tehran: Širkat-i Intišārāt-i ʿIlmī va Farhangī and Mīrāṯ-i Maktūb, 1375 [1996]) .

5 Šahmardān ibn Abī-al-Ḫayr Rāzī, *Rawżat al-munajjimīn*, ed. Jalīl Aḫavān-Zanjānī (Tehran: Mīraṯ-i Maktūb, 1382 [2003]).

6 Naṣīr-al-Dīn Ṭūsī, "Muḫtaṣar dar maʿrifat-i taqwīm," in *Safīna-yi Tabrīz, girdāvarī va ba ḫaṭṭ-i Abū-al-Majd Muḥammad ibn Masʿūd Tabrīzī, tārīḫ-i kitābat: 721-3 qamarī, čāp-i ʿaksī az rū-yi nusḫa-yi ḫaṭṭī-yi kitābḫāna-yi Majlis-i Šawrā-yi Islāmī*, ed. Nasrollah Pourjavady (Tehran: Markaz-i Našr-i Dānišgāhī, 1381 [2003]), 381–86.

7 Qumī, *Tarjuma-yi al-Madḫal* (note 4), III/1, 108; Ṭūṣī, "Muḫtaṣar dar maʿrifat-i taqwīm" (note 6), Section 8, 382; Kāšānī, I/5, fol. 365v.

8 Ṭūṣī, "Muḫtaṣar dar maʿrifat-i taqwīm" (note 6), Section 10, 382.

9 Qumī, *Tarjuma-yi al-Madḫal* (note 4), III/1, 103; Rāzī, *Rawżat al-munajjimīn* (note 5), II/2, 19; Ṭūṣī, "Muḫtaṣar dar maʿrifat-i taqwīm" (note 6), Section 2, 381; Kāšānī, I/4, fol. 365v.

10 At that time and place, the group of the seven planets included the five planets from Mercury to Saturn (only these planets could be observed from Earth by the naked eye) plus the Sun and the Moon, both of which were also considered to be planets rotating around Earth.

11 Both Rāzī and Ṭūṣī explicitly indicate the use of the *last* letters for abbreviations as *āḫarīn ḥarf* and *ḥarf-i āḫar*, accordingly (for exact page references, see note 12).

12 Rāzī, *Rawżat al-munajjimīn* (note 5), II/7, pp. 26–27; Ṭūṣī, "Muḫtaṣar dar maʿrifat-i taqwīm" (note 6), Section 7, p. 382; Kāšānī, I/7, fols. 365v–366r.

13 With regard to the word *hābiṭ* and its abbreviation by the *first* letter, mention should be made that the word *hubūṭ,* which derives from the same Arabic root and means "dejection," or the position at which a planet effects its minimal astrological influence, was abbreviated using its *last* letter *ṭa'*; this point will be discussed further in the essay.

14 Ṭūṣī, "Muḫtaṣar dar maʿrifat-i taqwīm" (note 6), Section 10, 382.

15 Rāzī, *Rawżat al-munajjimīn* (note 5), II/7, 26; Ṭūṣī, "Muḫtaṣar dar maʿrifat-i taqwīm" (note 6), Section 12, 383; Kāšānī, I/6, fol. 365v.

16 Ṭūṣī, "Muḫtaṣar dar maʿrifat-i taqwīm" (note 6), Section 12, 383.

17 Ibid.

18 Ṭūṣī, "Muḫtaṣar dar maʿrifat-i taqwīm" (note 6), Section 15, 383.

19 Kāšānī, I/8, fol. 366r.

20 Rāzī, *Rawżat al-munajjimīn* (note 5), II/7, 26–27; Ṭūṣī, "Muḫtaṣar dar maᶜrifat-i taqwīm" (note 6), Sections 12–13, 383; Kāšānī, I/8, fol. 366r.

21 See samples in Gacek, "Abbreviations" (note 2), 3–4.

22 Ṭūṣī, "Muḫtaṣar dar maᶜrifat-i taqwīm" (note 6), Section 15, 383.

23 Rāzī, *Rawżat al-munajjimīn* (note 5), II/7, 27.

24 ᶜAlī-Akbar Dihḫudā, *Luġatnāma*, ed. Muḥammad Muᶜīn and Jaᶜfar Šahīdī (Tehran: Mu'assisa-yi Intišārāt va Čāp-i Dānišgāh-i Tihrān, 1372–73 [1993–94]); Majd-al-Dīn Muḥammad ibn Yaᶜqūb al-Fīrūzābādī and Naṣr al-Hūrīnī, *Al-Qāmūs al-Muḥīṭ*, 4 vols. (Cairo: Al-Maṭbaᶜah al-Miṣriyyah, 1925); Gacek, "Abbreviations" (note 2), 3; Salih H. Alić, "Problem kratica u arapskim rukopisima (so spiskom arapskih kratica iz 16. vijeka)" [Problem of abbreviations in Arabic manuscripts], in *Prilozi za orijentalnu filologiju XXVI (1976)*, ed. Sulejman Grozdanić, 199–212 (Sarajevo: 1978). These abbreviations are mentioned in the *Dictionary* (*Luġatnāma*) of ᶜAlī-Akbar Dihḫudā. I used the CD-ROM and online versions of it; see the entries about corresponding separate letters, as well as the introduction to Fīrūzābādī, *Al-Qāmūs al-Muḥīṭ*; cf. also Gacek, "Abbreviations" (note 2), 3. It is worth noting that in modern editions of *Al-Qāmūs al-Muḥīṭ*, the abbreviation for the word *qaryah* is given as *tā' marbūṭah*, while Naṣr al-Hūrīnī (the fifteenth-century CE commentator on *Al-Qāmūs al-Muḥīṭ*) explicitly states in a poem that the word *qaryah* is in fact abbreviated by means of the Arabic letter *hā'*. I learned about this poem from the article by Salih H. Alić (Alić, "Problem," 206).

25 Kāšānī, I/8, fol. 366r.

26 Rāzī, *Rawżat al-munajjimīn* (note 5), II/7, p. 26.

27 Adam Gacek, "Ownership Statements and Seals in Arabic Manuscripts," *Manuscripts of the Middle East* 2 (1987): 89; Gacek, "Abbreviations" (note 2), 3; Gacek, *Arabic Manuscripts: A Vademecum for Readers* (Leiden: E. J. Brill, 2009), 85.

28 Māmaqānī, *Muᶜjam* (note 3), 92–95.

29 Ismāᶜīl ibn Ḥammād Jawharī, *Muᶜjam al-Ṣiḥāḥ: Qāmūs ᶜArabī - ᶜArabī murattab tartīban alifbā'iyyan wafqa awā'il al-ḥurūf*, ed. Ḫalīl Ma'mūn Šīḥā (Beirut: Dār al-Maᶜrifah, 1426 [2005]). See the editor's introduction (p. 17), where this rearrangement and the reasons for it are discussed.

CHAPTER 9

Creating a Codicology of Central Asian Manuscripts

Susan Whitfield

Whereas paleography and codicology are mature disciplines in Western manuscript studies, in many Asian traditions they are barely recognized. Indeed, the contrast goes further than this: many of the Asian traditions do not differentiate between manuscript and textual studies. This is especially the case in those cultures where texts on manuscripts have been copied repeatedly and the old manuscript discarded: the worth was considered to lie entirely in the writing and not in the object. While in certain cultures—for example, in China and in many Buddhist cultures—the written word was considered sacred and the cultural norm was for preservation, the emphasis was still primarily on the text, to a lesser extent on the paleography, but hardly at all on the object.[1]

This essay considers an extensive collection of manuscripts from these cultures and one that bridges the classical and medieval periods. The manuscripts are those excavated from archaeological sites in the ancient Silk Road kingdoms of the Tarim Basin in Chinese Central Asia and from a hidden library cave to the east in the Gobi. As a result of archaeological expeditions funded mainly by imperial powers during the early twentieth century, these manuscripts are now dispersed among collections in Europe and Asia.[2] Fortunately, most were deposited and remain in public institutions. The International Dunhuang Project (IDP), founded by the major holding institutions in 1994, has been leading efforts to conserve, catalogue, and digitize them over the past decade, and, as a result, by 2015 details of more than a hundred thousand manuscripts could be viewed freely online.[3]

This essay first gives an overview of the collections and their characteristics. It discusses their potential for understanding the local, regional, and transregional development of manuscript cultures throughout the Silk Road period. It ends with a consideration of the potential of using digital data to understand this transmission.

The Central Asian Collections

The Central Asian Collections number more than 150,000 manuscripts, many fragmentary, with more being discovered. Apart from the finds in the Dunhuang Library Cave, the manuscripts were found in a variety of archaeological contexts dating from the Silk Road period, in military forts, Buddhist stupas, residential and official buildings, rubbish heaps, and tombs. They date from the first century BCE to the fourteenth century CE.[4]

The Dunhuang Library Cave, generally accepted as a Buddhist repository, could be compared to the Cairo Genizah except that, along with the Tarim Basin finds, it is differentiated by a greater diversity of languages, scripts, media, and formats of its contents.[5] Texts exist written in over twenty languages and scripts, including those of the empires and kingdoms on the periphery of Central Asia during this period—Chinese, Tibetan, Iranian, Indian, and Turkic languages—but also those of Central Asian kingdoms—including Sogdian, Khotanese, Gandhāran, and Tocharian.[6] There are also manuscripts suggesting the existence of long-distance traders or travelers, such as Greek and Judeo-Persian (Persian in Hebrew script). Scripts are used and reused by different languages: Syriac, a script derived from Aramaic, for example, was first borrowed by the Sogdians in the fourth century to transcribe their Middle Iranian language. Sogdian script was then adapted by the Uygurs for their Old Turkic language when they moved into the Tarim Basin in the mid-ninth century. The Uygur script was used by the Mongolians from the early thirteenth century. There are manuscripts representing all these combinations.

In terms of content, many of the manuscripts found are not part of Buddhist tradition but contain secular texts, on literary, military, legal, official, scientific, and medical matters. There are also texts from the Manichaean, Daoist, Zoroastrian, and Nestorian Christian religious traditions.[7]

A large number of the manuscripts are on paper, especially those from Dunhuang, an area linked more closely than its Tarim Basin neighbors to the Chinese world, where paper had been invented by the third or second century BCE. By the third and fourth centuries, papermaking had matured, and commonly used materials in central China included pulp made from various materials, including hemp (*Cannabis sp.*), ramie *(Boehmeria nivea)*, and the bark of the paper mulberry tree (*Broussonetia sp.*).[8] As comments in the documents in Dunhuang reveal, paper imported from central China was expensive and not always readily available. Local papermaking developed and started spreading westward over the following centuries. These used locally available materials, including paper mulberry and hemp in the Tarim but other species as well. So, for example, a study analyzing the paper of Tibetan manuscripts from Dunhuang shows a group containing *Thymelaeaceae* family plant fibers (specifically *Daphne* or *Edgeworthia sp.*); this paper was probably made on the Tibetan plateau, where this species is commonly available.[9]

However, the earliest Chinese Central Asian manuscripts are on wood, the predominant support in China before paper, and are found in the Chinese-built defensive walls and forts to the north of Dunhuang dating from the start of the Silk Road, the first century BCE onward (Figure 9.1).[10] It is hypothesized that the wood is from the poplar species, commonly found in the oases of the Gobi and Taklamakan deserts and in widespread use as a building material.[11]

The first- to fourth-century kingdom of Kroraina in the Lop and east Taklamakan also used wood to transcribe official documents created as an archival record (Figure 9.2). Paper was probably not readily available at this time, but the distinctive forms of the manuscript envelopes—quite unlike the Chinese leaves made from long and narrow wood slips strung together—also suggest that manuscript influences were not from China. The written language of the kingdom was a locally influenced Gandhāran dialect of Prakrit, and the texts were written in the Kharoṣṭhī script, used only in Central Asia during these few centuries. Scholarly discussion

Figure 9.1. Early examples of Chinese Central Asian manuscripts on strips of wood (Or.8211/26, 28, 29, 30, 31, 32, 35).

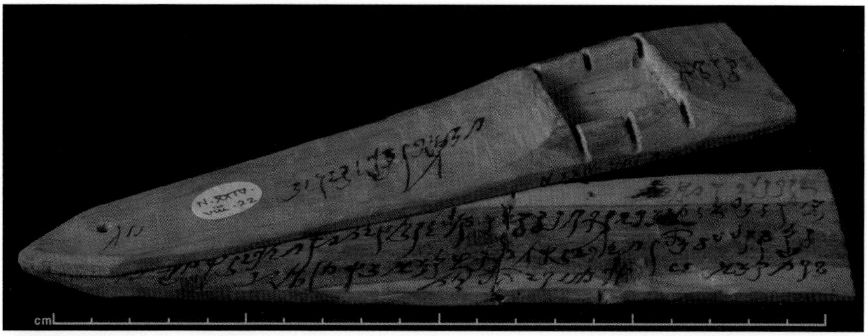

Figure 9.2. Example of a document inscribed on wood from the kingdom of Kroraina (Or.8211/1414).

continues about whether the Kushan empire extended into the Tarim and, if so, how far, but its influence was certainly felt, as evidenced by the bilingual Chinese-Gandhāran coins issued in Khotan during this period.[12] The Kushan empire would have formed a conduit for influences from Iran and India, but the extant manuscripts in Gandhāran are on birch bark. Contemporary wooden manuscripts written in carbon ink (rather than inscribed with a stylus) have been discovered elsewhere, including from the Vindolanda Roman fort near Hadrian's Wall in northern England.[13]

The use of wood continued even after paper became available. In some cases, this was almost certainly owing to availability, cost, and appropriateness of use: for example, when soldiers of the Tibetan empire moved into the southern Tarim following the retreat of the Chinese in the mid-eighth century, they used wood slips at their military forts to record everyday orders and tallies. Tibetan Buddhist texts, however, are generally written on paper.

Manuscripts are also found on other media possibly locally obtained, such as leather, and on media certainly obtained from elsewhere, such as palm leaves and birch bark as used in India and the mountain lands to its north, and bamboo and silk, probably from China. However, paper and wood remain the dominant media.

Figure 9.3. Sealed wooden envelope (Or.9268).

The formats of the manuscripts show great diversity. Some, such as the wood slips, clearly belong to an existing and known tradition—the Chinese classical period. But others, such as the Krorainian manuscripts, have no obvious predecessors. Although some Bactrian manuscripts are on wood, they are not in a similar form; they are, however, similarly fastened with string and clay seals.[14] Rectangular rather than wedge-shaped, sealed wooden envelope documents are found later in Khotan, the neighboring kingdom to the west, and were perhaps influenced by Kroraina, but there is little other evidence of any legacy of this unusual format (Figure 9.3).

The few paper fragments from Kroraina carry Chinese text, a clear link with the Chinese tradition. They are small fragments, and it is difficult to guess their original format.[15]

Apart from small fragments of paper, the earliest complete manuscripts on paper are five folded sheets found by the defensive walls north of Dunhuang. They contained letters written in 313–14 CE by Sogdians, the merchants of the eastern Silk Road whose homelands were the city-states of present-day Samarkand, Bukhara, and Khiva but who were based in the Hexi corridor leading from Dunhuang into China. These letters are simple paper panels that contain the letter text on one side and are then folded; the address and recipient were written on the outside (Figures 9.4A and B).[16] Aurel Stein noted that their size corresponds to the height of contemporary Chinese wood slips.[17]

The use of single sheets continued (in a Judeo-Persian document concerning the trade of sheep, for example), but two formats dominated for the following few centuries: the *pothi* and the scroll, the latter used mainly for Chinese texts and the former found in other languages, including Tibetan, Sanskrit, Khotanese, and Tocharian.

Pothi or *pustaka* refers to the shape of palm leaves used for manuscripts in India. Although few palm-leaf manuscripts survive from this period—the

---

Following spread:
Figure 9.4. A Sogdian letter (top: recto; bottom: verso) (Or.8212/92.1).

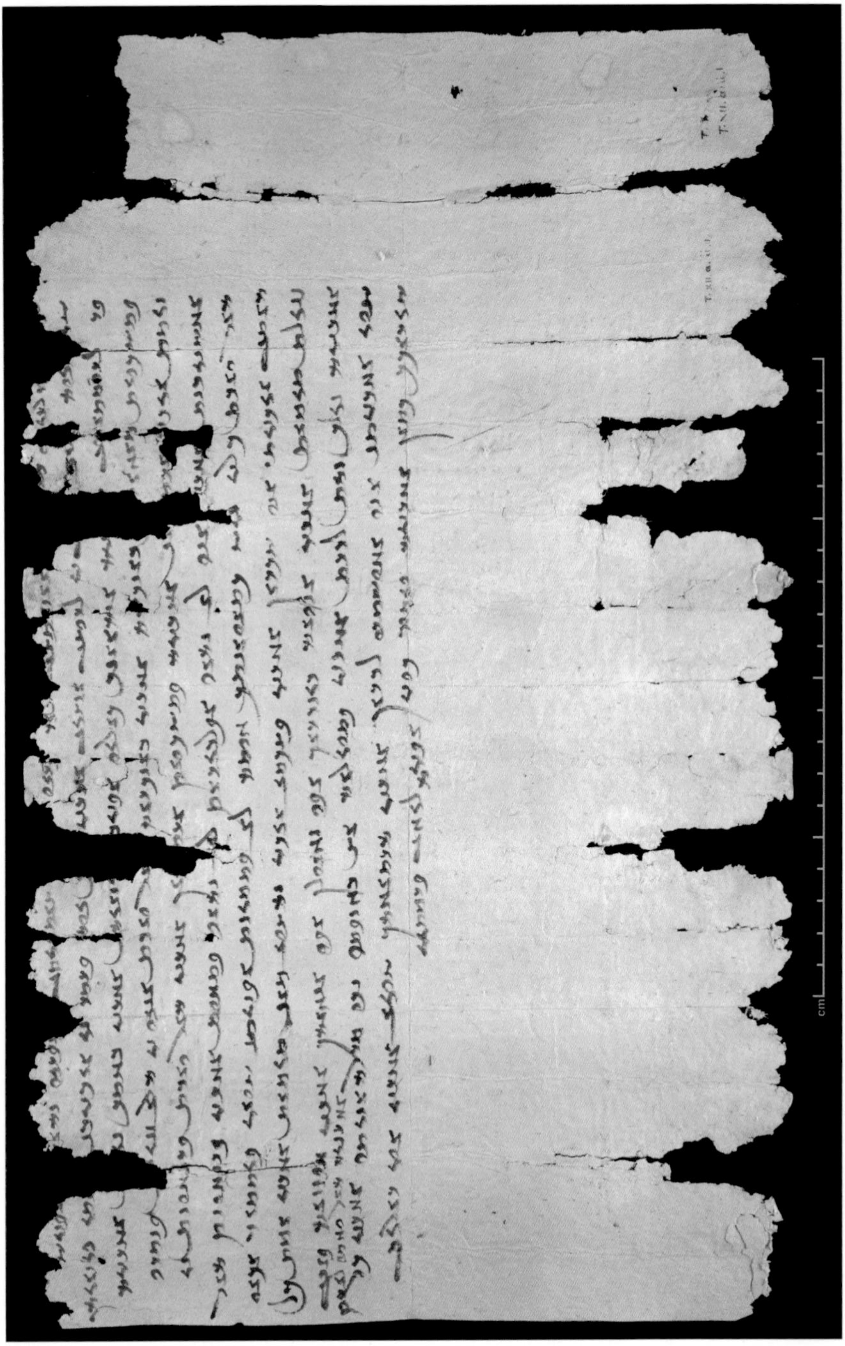

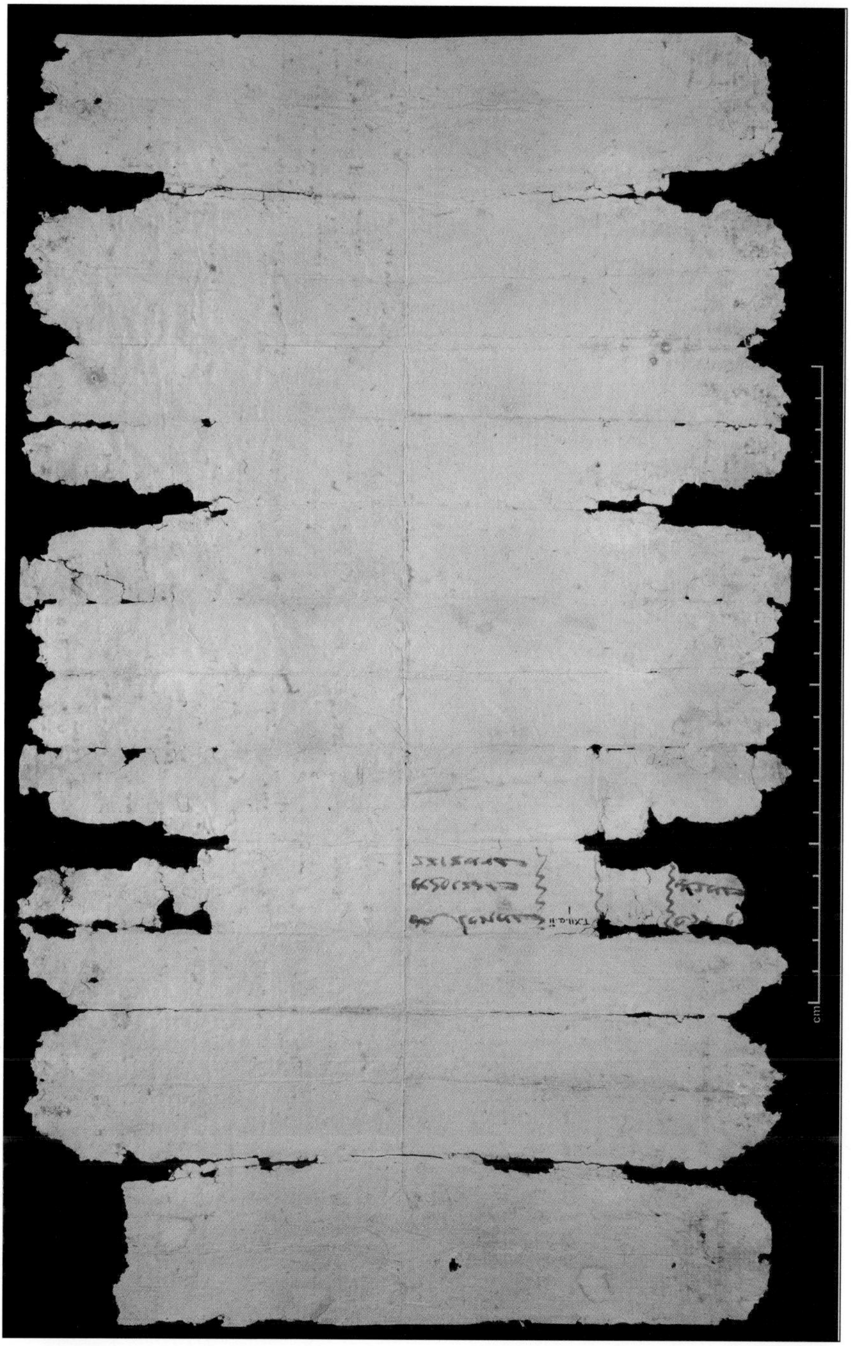

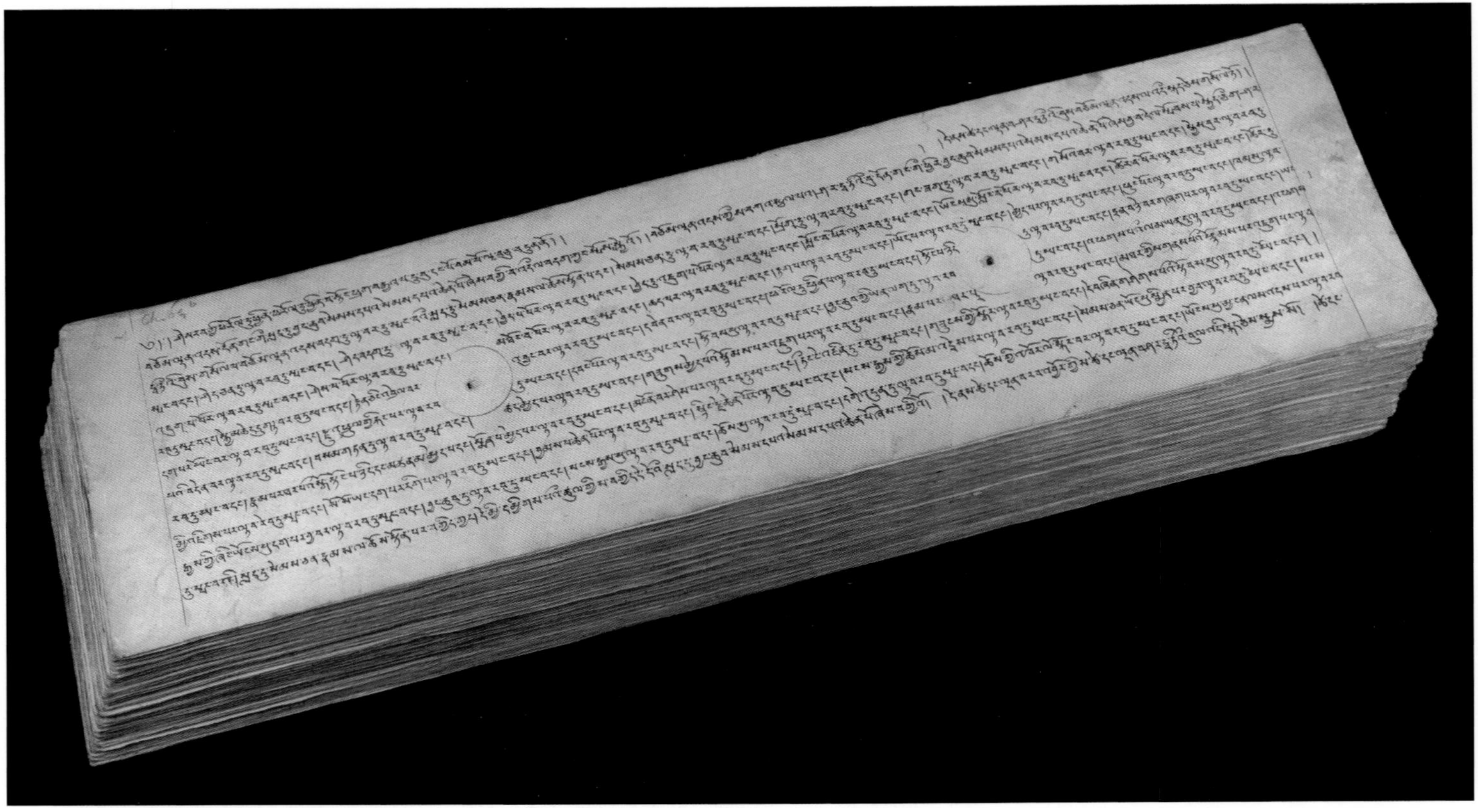

Figure 9.5. Paper *pothi* (IOL Tib J 105).

ones found in Dunhuang are exceptional—this tradition is clearly attested and continues today. In Tibet and Central Asia, where palm leaves were not available but there were local papermaking industries, the Indian form was retained, but the use of paper meant that there was more flexibility in the size of the *pothi*. Palm-leaf *pothi* vary in width; the ones from Dunhuang are long at 55 centimeters but are usually only about 5 centimeters high, limited by the size of the leaf. The paper *pothi,* especially the later Tibetan ones found at Dunhuang, are up to 10 centimeters high, doubling the size of the palm leaves (Figure 9.5).[18] Chinese texts on *pothi* also exist, notably at Toyuk near Turfan and at Dunhuang, but they are usually in vertical format, with margins and guidelines, as seen in scrolls, and most have a string hole. At least one transcends the dictates of the original form, being almost square, but still retains the string hole in the center (Figure 9.6).[19]

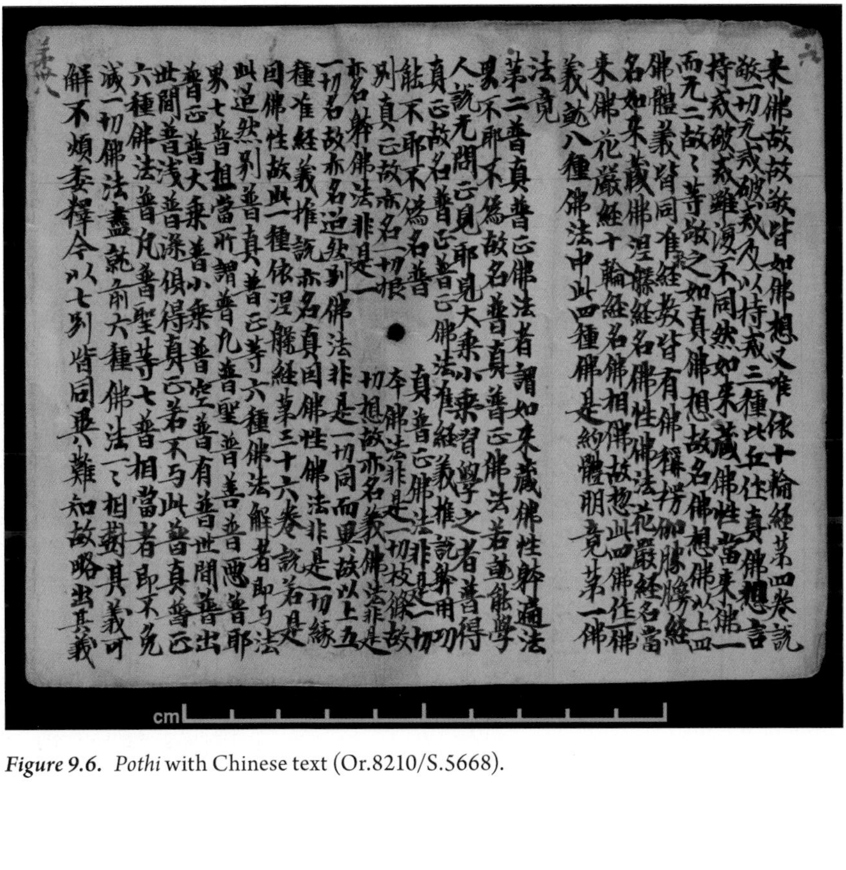

Figure 9.6. *Pothi* with Chinese text (Or.8210/S.5668).

It is only a step from the *pothi* to the concertina form, where the *pothi* were joined together to form a folding book, and most are in Tibetan.[20] A few have been found in Chinese, but perhaps the most interesting is a bilingual manuscript containing Chinese and Tibetan.[21] The Chinese reads vertically—as in scrolls. The Tibetan text, a commentary, is read horizontally, the manuscript then resembling the usual *pothi* (Figure 9.7). There is a redundant string hole, retaining this feature of the *pothi* form, even though it was no longer necessary to hold the leaves together, as they were glued at their edges.

Chinese paper scrolls have been generally placed in a direct lineage from the rolled-up wood slips found in central China, but there is also the possible influence of the separate tradition of birch-bark scrolls used in Gandhāra, themselves influenced by Hellenistic traditions.[22] Several collections of Buddhist birch-bark manuscripts dating from the first century have come into the market over the past three decades and are now held at the British Library, in the Senior Collection at the University of Washington, and in the Schøyen Collection.[23] As the earliest Buddhist manuscripts yet found, they have been very important for scholarship. They are in scroll format, usually composed of more than one sheet of bark sewn together. The material is extremely friable, making conservation difficult, and, even when unrolled and "flattened," it is often impossible to separate every layer. They are written in Gandhāran Prakrit in Kharoṣṭhī script. The discovery of these contextualized a similar manuscript, a birch-bark scroll in Gandhāran, acquired in 1892 near Khotan by the French explorers MM. Dutreuil de Rhins and Grenard.[24] Only a few other fragmentary birch-bark manuscripts have been found in this region with Gandhāran in Kharoṣṭhī and none from Kroraina, which might be expected to have had strong links with the Gandhāran region in the first to third century. However, the Krorainian manuscripts come mainly from a few non-Buddhist administrative archives, and the lack of other discoveries does not mean they did not once exist in this community.

Figure 9.7. *Pothi* with Tibetan text (Or.8210/S.5603).

所相攝受計著者由心量故使起我見攝受計著矣不覺自心現量於智尒炎而
起妄想者由不覺心計有境智起妄想矣從此第五依於斷見其義者何　經云妄想
故外住非住觀察不得依於斷見　解曰妄想故外住非住觀察不得依於斷見者由妄
想故見有外境於外非住不能觀察名為不得不知真智真室不得便起斷見謗實智矣
從此偈誦其義者何　經云尒時世尊欲重宣此義而說偈言　有諸攀緣事
智慧不觀察　此无智非智　是妄想者說　於不異相住　智慧不觀察　障礙及遠近
是名為耶智　老小諸音實　而智慧不生　而實有尒炎　是亦說耶智　解曰初之一行
上之三句誦大慧難有攀緣事智不觀察名為无智是智矣其第四句仏言妄想愚癡之
者作如是說非真智矣後之兩行誦上五難於不異相是誦初難後行尒炎誦第二難餘義

可知從此大文第六明說宗通以承前文了假施設入无所有故須了悟說通宗通故次明矣於
中有五一標迷二通二大慧請問三仏許為說四大慧受教五世尊正解且初第一標迷二通其
義者何　經云復次大慧愚癡凡夫无始虛偽惡邪妄想之所迴〻轉〻時自宗通及說通
不善了知著自心現外性相故著方便說於自宗四句清淨通相不善分別　解曰愚癡凡
夫妄想迴轉見心外境名　為迴轉心迴轉時迷於二通不善了知自心現境著方便說二乘
之流見心外境不知自心名　為自宗心之自宗離於四句清淨通相不能善解故有迷矣
從此第二大慧請問其義者何　經云大慧白仏言誠如尊教唯願世尊為我分別說
通及宗通我及餘菩薩摩訶薩善於二通來世凡夫聲聞緣覺不得其短
解曰大慧請說若善二通人不得短以宗說通離見過故不得短矣從此第三仏許為說

The British Library Kharoṣṭhī collection includes a single birch-bark manuscript in Sanskrit with Brāhmī script.[25] Other Sanskrit manuscripts have also been found in the region, including from Bamiyan, possibly dating from as early as the second century onward, and near Taxila, probably dated to the fifth century, the latter found, like the early collection, in a clay pot. Sanskrit birch-bark manuscripts are more frequently found in eastern Central Asia, most especially on the northern Silk Road. The first was an acquisition by Lieutenant Bower in 1890 near Kucha and now in the Bodleian Library, Oxford.[26] This is in *pothi* format, and, although many of the other discoveries made since are small fragments, making it impossible to surmise the original format, a few clearly were *pothi*.[27]

From the ninth and tenth centuries, we start to see examples of early codices, almost all in paper, but containing Chinese, Tibetan, Turkic, and other scripts. Some of the early forms are a clear single step from the concertina, including a printed document, Or.8210/P.11, where each folio is folded back on itself to form a folio of two pages with the blank sides inward and pasted. Thus the concertina is restricted to a single opening. The wrapped back binding is similar and ideally suited to printing, but the original folios are not glued together in the first place; rather, they are tied at their unfolded edges to form a booklet.[28] In a few cases these are *pothi* shaped, but in most cases, the pages are more of a rectangle (Figure 9.8).[29]

Other forms also appear over this period: butterfly, stitched, and, still controversial, whirlwind.[30] The codex becomes the dominant form in central China as woodblock printing becomes widespread over the next few centuries, with the wrapped-back form becoming the standard, with each woodblock used to print one folio, making two pages folded back.

Around this time another culture emerges stretching from Dunhuang north to the borders of what is now Mongolia: the Tanguts. A large cache of manuscripts and printed documents found in a stupa in the northern border city of Karakhoto shows the influence of Chinese and Tibetan manuscript and printing traditions. There are a few scrolls, *pothi*, and concertinas, but the largest number are in codex format and include many printed items.[31] The paper, according to analysis of the Russian collection, seems to be of a distinctive local production.[32]

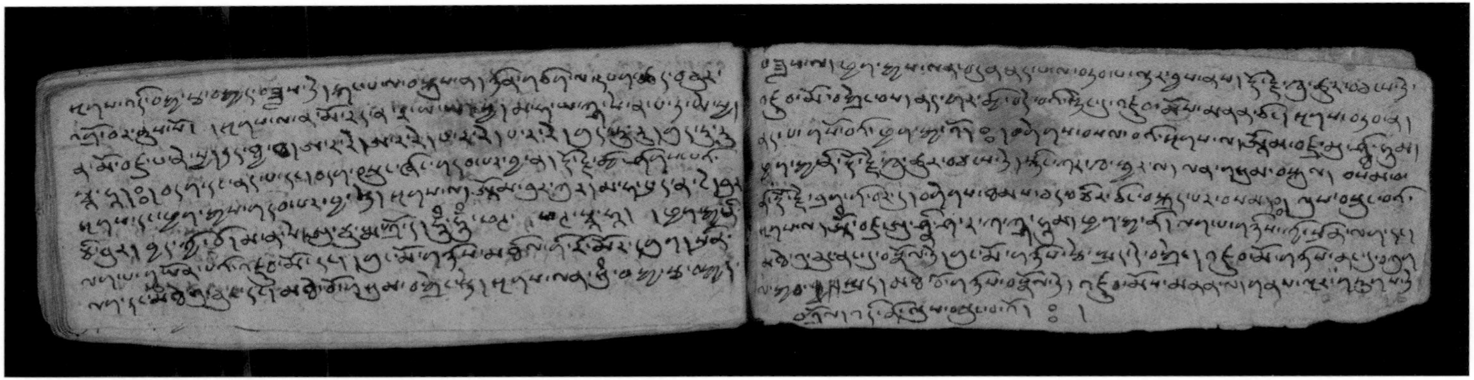

Figure 9.8. Example of early codex form (IOL Tib J 401).

This summary is by no means exhaustive. Other transmissions have been suggested as, for example, the influence of Near Eastern–Syriac on Manichaean formats, most especially the relative orientation of the text and images.[33] This period also saw the appearance of Buddhist texts written in gold and silver ink on indigo-dyed paper. At around the same time but at the other end of Eurasia, we see the Qur'an written in gold and silver ink on indigo-dyed parchment. Is this a case of independent developments or transmission? We have much work yet to do before we can answer this question, but making available a critical mass of data is an essential first step.[34]

Central Asian Manuscript Cultures and Interactions

Despite the size of the Central Asian collections, their variety and distribution confound easy classification. This is compounded by the lack of historical context: the first history of the Taklamakan kingdoms was published only in 2012, and there is yet to be a monograph on any single kingdom.[35] The most studied manuscripts are those from the Dunhuang Library Cave, but understanding of this collection as a whole was impeded by the distribution of the cave's contents to institutions in Europe and Asia within fifteen years of its discovery. It was only when institutions started making microfilms available beginning in the 1950s that scholars were able to study the collection.[36] Over the past decade, the digitization of the manuscripts and their free availability online on IDP has greatly increased the possibilities for research.

From the above summary of the *pothi* and scroll formats, enabled by being able to search across large parts of the Central Asian collections on IDP, we might hypothesize that two regional manuscript traditions, one in India and the other in Gandhāra, developed about the first century CE to transcribe Buddhist texts, and the format was molded by the available materials—palm leaves and birch bark—but also, in the latter case, possibly by the scroll tradition from further west. Some of these manuscripts were carried across the mountains to the Tarim Basin kingdoms; some, although this is speculation, may have found their way to China or at least to its western borders.

Over the next few centuries, with the demise of the Kharoṣṭhī script, the *pothi* form became preferable in the Central Asian tradition, whether on paper, birch bark, leather, wood, or silk. The form was borrowed by several other cultures to produce manuscripts in Tocharian, Khotanese, and Tibetan. The paper scroll became the dominant form for Chinese manuscripts, both secular and religious. Possibly during the Tibetan era and afterward (mid-eighth century), when the links with central China were probably weaker, the *pothi* was also used for Chinese texts.

While the above is a hypothesis, it has become possible to try to discern patterns in the extant manuscript corpus owing to its much greater availability over the past decade. Now that images and metadata of much of the material are freely available online, the next step is to start using the data to test such hypotheses but also to encourage research to generate more data.

Manuscripts are not neutral carriers of written cultures; they are objects. To understand them, we need to understand the societies that created and used them, while the manuscripts themselves will also shed light on the culture, its influences, and interactions. Historical, textual studies, and codicology in this area are all in their infancy, with few detailed studies, for example, on the transmissions, translations, and variations of various Buddhist texts between cultures or on the transmission and development of manuscript media and formats.

Some of the manuscripts found in an archaeological context have had several functions. For example, many of the manuscript fragments found in Buddhist stupas were probably votive offerings.[37] The integrity of the text or the object was not important—fragments contained words of the Buddha, and even a single word was valued. But this veneration for paper containing words was not restricted to Buddhism: it was also part of the Chinese tradition. A ninth-century Chinese writer complained that manuscripts were being used as toilet paper; this is one possible explanation for the scattered discovery of a Chinese manuscript found in eastern Kroraina. It had been torn into several pieces, and the pieces were found at four different archaeological sites. They were many miles apart but, as Stein points out, lying on a straight line that probably contained settlements on the ancient route through this kingdom.[38]

The practice needs to be understood, or otherwise we might draw misleading conclusions. For example, a *terminus ante quem* is given for a manuscript found at the Endere temple in Kroraina based on the date this temple was abandoned.[39] But there is evidence that temples continued to be used for offerings even when abandoned; they were still holy places, and the pious would make detours to visit them. There is a possibility, therefore, that the manuscript could have been placed in the temple after the settlement was deserted, and so it is not in itself sufficient evidence for dating.

A significant number of manuscripts have been recycled for practical purposes, to form the lining for soles of shoes, as repair patches on other manuscripts, or, in the case of Tibetan wood slips from military forts, refashioned to become spatulas, spoons, and scrapers once their use as textual holders was over.[40]

All these are factors that need to be taken into account when working with this material.

Digital Data and the Future

IDP went online in 1998 with images of a few thousand manuscripts. In 2014 it offered free access to more than 450,000 images of more than 100,000 manuscripts from more than 20 collections worldwide, with more being added daily. As well as metadata and catalogues (and where possible, transcriptions and translations of the text), it has always sought to place the manuscripts in their archaeological context. IDP's remit is to record all the archaeological material, not just manuscripts, including paintings, textiles, and artifacts, as well as information about the sites themselves.

Over the past decade, IDP has also started to collate and record paleographical and codicological data. Initial research revealed that while there had been several attempts in the past to research and publish such information, the samples of manuscripts under scrutiny had varied from a handful to a few score. In such a diverse and widely dated group of manuscripts, a sample of twenty will have no significant statistical application to the whole.

In addition, new research did not always build on or refer to previous work, usually because of lack of access.[41]

Testing on the paper started soon after the manuscripts were found. M. Aurel Stein (1862–1943), for example, sent several fragments that he found to Julius R. von Wiesner (1838–1916), professor of botany at the University of Vienna.[42] A further group of samples from the Stein collection, manuscripts from Dunhuang, was tested by Robert H. Clapperton in the 1930s and cited in his 1934 book on papermaking. T. T. Tsien's seminal work, *Written on Bamboo and Silk*, appeared in 1962, making Chinese historical sources on papermaking and bookbinding widely available.[43] Improved microscopic techniques were applied to a slightly larger group of the Dunhuang manuscripts in the Stein collection by T. J. Collings and W. D. Milner.[44] Their results, published in 1979, did not tally entirely with Clapperton's findings, but this is not surprising given the improvement in technology and the difficulty of identifying some plant fibers when used in paper.

The new discoveries of early paper fragments in Chinese central Asia by Chinese archaeologists from the 1930s onward became the subject of study from around this period, with Pan Jixing publishing an English overview of his work in 1981 for the International Association of Paper Historians.[45] New finds have prompted more study in China, summarized in a second edition of T. T. Tsien's work published in 2004.

Akira Fujieda was among the first to approach the manuscripts as a collection, concentrating on the Dunhuang Library but also looking at papers from Turfan. His codicological classification remains the standard, although it has yet to be widely applied or developed further.[46] J. P. Drège was another pioneer in the study of book formats and collation of codicological information, who also concentrated his attention on the Dunhuang Library.[47] The paper conservator Anna-Grethe Rischel made studies of larger sample sets, carrying out macroscopic analysis of papers in Danish, Swedish, and British collections.[48]

The conference on Dunhuang manuscript forgeries organized by IDP at the British Library in 1997 brought these scholars together for the first time and was intended as a springboard for future collaborative research to develop these initial studies. Since then, IDP has concentrated on ensuring that all the available data about the Dunhuang and other Central Asian manuscript collections are input into its database. This process is ongoing and dependent on external grants, but, over the past five years, with a critical mass of data available, IDP has also started to coordinate systematic studies working with paper scientists and others. Projects include fiber analysis with Agnieszka Helman-Wazny, paper morphology with a team from Ryukoku University, and pigment analysis with scientists from London and Berlin, collating all existing data and carrying out new mass-sampling where data are lacking. The ensuing open data set will be available for use by scholars worldwide. Linked with historical, archaeological, and other data, it will offer a powerful tool for researchers. It is hoped that it will enable them to start to map and understand the different traditions and complex transmissions of Central Asian manuscripts.

Notes

1 In many cultures, it was considered perfectly acceptable to make copies of manuscripts—thus preserving the text—and then to discard the old manuscript or object.

2 Summaries of the expeditions and collections can be found on the IDP collection pages, available at http://idp.bl.uk/pages/collections.a4d. Chinese archaeologists continue to uncover more manuscripts from the same and related sites, and these are also part of the IDP.

3 All the manuscripts mentioned in this chapter can be viewed on the IDP Web site at http://idp.bl.uk through a simple search using the manuscript number as given.

4 For a representative selection, see Susan Whitfield, ed., *The Silk Road: Trade, Travel, War and Faith* (London: British Library, 2004). (London: British Library, 2004).

5 See Sam van Schaik and Imre Galambos, *Manuscripts and Travellers: The Sino-Tibetan Documents of a Tenth-Century Buddhist Pilgrim* (Berlin: Walter de Gruyter, 2012), 18ff, for a summary of the various theories for the creation and sealing of the Dunhuang Library Cave.

6 See the map on pp. 20–21 of *Turfan Studies* (Berlin: Berlin-Brandenburg Academy of Sciences and Humanities, 2007), downloadable at http://turfan.bbaw.de/bilder/Turfan_engl_07.pdf.

7 Many of the sites where manuscripts were found were pre-Islamic, such as Kroraina, or were Buddhist contexts, such as the Library Cave or stupa deposits.

8 Note that this is a different species from the mulberry tree used in sericulture (*Morus* sp., especially *Morus alba* or white mulberry).

9 A. Helman-Ważny and S. Van Schaik, "Witnesses for Tibetan Craftsmanship: Bringing Together Paper Analysis, Palaeography and Codicology in the Examination of the Earliest Tibetan Manuscripts," *Archaeometry* 55, no. 4 (2013): 707–41.

10 Bamboo was also used in China but would not have been readily available here.

11 Rachel Roberts, "House Building in Ancient Niya," 39 (Spring 2012): 4–5; available at http://idp.bl.uk/archives/news39/idpnews_39.a4d#3.

12 J. Cribb, "The Sino-Kharoṣṭhī Coins of Khotan—Their Attribution and Relevance to Kushan Chronology, Parts 1–2," *Numismatic Chronicle* 144 (1984): 128–52; 145 (1985): 136–69. Dates given in this article have been revised to ca. 30–150 CE following more recent finds. See also Vidula Jayaswal, ed., *Glory of the Kushans: Recent Discoveries and Interpretations* (New Delhi: Aryan Books International, 2012).

13 Vindolanda Tablets Online, http://vindolanda.csad.ox.ac.uk/.

14 See an example on the IDP Web site at http://idp.bl.uk/database/oo_loader.a4d?pm=SF2004/16.

15 See, for example, the British Library manuscript Or.8212/492. One fragment appears to be part of a scroll (Or.8212/511). In his 1953 catalogue, Henri Maspero points out that a variant character suggests that this is later, that is, from the seventh century onward, but the writing, the paper, and its provenance suggest an early fourth-century date, in keeping with the date of Kroraina (Maspero, *Les documents chinois de la troisième expédition de Sir Aurel Stein en Asie Centrale* [London: British Museum, 1953], 79, no. 253).

16 London, British Library, Or.8212/92-8. See Nicholas Sims-Williams, "The Sogdian Fragments in the British Library" *Indo-Iranian Journal* 18 (1976): 43–82; and, for translations, "The Sogdian Ancient Letters," trans. Nicholas Sims-Williams, University of London, http://depts.washington.edu/silkroad/texts/sogdlet.html.

17 Aurel Stein, *Serindia* (Oxford: Clarendon, 1921), 672.

18 Paper size was limited by the size of the papermaking mold.

19 This is a Buddhist work in Chinese: London, British Library, Or.8210/S.5668.

20 Jean-Pierre Drège, "Les accordéons de Dunhuang," in *Contributions aux études de Touen-houang* 3, 195–204 (Paris: École française d'Extrême-Orient, 1984).

21 London, British Library, Or.8210/S.5603.

22 Richard Salomon, with contributions by Raymond Allchin and Mark Barnard, *Ancient Buddhist Scrolls from Gandhāra: The British Library Kharosthī Fragments* (Seattle: University of Washington Press; London: British Library, 1999), 101.

23 See ibid. for an overview of the British Library collection; Mark Allon, *Ancient Buddhist Scrolls from Gandhāra: The Senior Collection* (Seattle: University of Washington Press, forthcoming) for an overview of the Senior Collection; thttp://www.schoyencollection.com/buddhismIntro.html for the Schøyen Collection.

24 H. W. Bailey, "The Khotan Dharmapada," *Bulletin of the School of Oriental and African Studies* 11, no. 3 (1945): 488–512.

25 Fragment 6, containing a medical text. As Salomon points out (*Ancient Buddhist Scrolls from Gandhāra* [note 22], 39), it is on a narrower scroll than the others, suggesting a different provenance.

26 A. F. Rudolf Hoernle, *The Bower Manuscript* (Bombay: British India Press, 1914) (repr. with additions, from vol. 22 of the New Imperial Series of the Archaeological Survey of India).

27 Berlin, Staatsbibliothek, SHT 14/2.

28 For a summary of types of bookbindings among Dunhuang manuscripts, see Jean-Pierre Drège, "Les cahiers des manuscrits de Touen-houang," in *Contributions aux études de Touen-houang*, vol. 1, 17–28 (Geneva and Paris: Librarie Droz, 1979). See also IDP, "Bookbinding," at http://idp.bl.uk/education/bookbinding/bookbinding.a4d; and Whitfield, *The Silk Road* (note 4), cat. nos. 257–60.

29 London, British Library, IOL Tib J 401, 510.

30 See Whitfield, *The Silk Road* (note 4), 298, no. 257.

31 Mikhail Piotrovsky, ed., *Lost Empire of the Silk Road: Buddhist Art from Khara Khoto (X–XIIIth Century)* (Milan: Thyssen-Bornemisza Foundation, 1993), 257–69.

32 "The paper is of local Tangut manufacture. Paper of this type has been found to be made of flax half stuff, with the addition of hempen fibre" (Piotrovsky, *Lost Empire of the Silk Road* [note 31], no. 75).

33 Zsuzsanna Gulacsi, "Manichaean Book Art," in Whitfield, *The Silk Road* (note 4), 212–13.

34 See François Deroche, *The Abbasid Tradition: Qur'ans of the 8th to the 10th Centuries AD* (London: Nour Foundation, with Azimuth Editions and Oxford University Press, 1992); and "Folio from the Blue Qur'an [Probably North Africa (Tu-

nisia)] (2004.88)," in *Heilbrunn Timeline of Art History* (New York: Metropolitan Museum of Art, 2000), available at metmuseum.org/toah/works-of-art/2004.88 (October 2008).

35 Valerie Hansen, *The Silk Road: A New History* (Oxford: Oxford University Press, 2012).

36 Lionel Giles, cataloguer of the Stein collection at the British Library, produced an overview of the Dunhuang manuscripts in a 1941 lecture for the China Society: "Six Centuries at Tunhuang," published by the society in 1944. But one of the first to take advantage of the microfilms was Professor Fujieda Akira. His two-part overview, "The Tunhuang Manuscripts: A General Description," appeared in the Japanese journal *Zinbun* 9 (1966): 1–32, and 10 (1969): 17–39.

37 Salomon, *Ancient Buddhist Scrolls from Gandhāra* (note 22), 30–31, concerning the interring of Buddhist manuscripts.

38 London, British Library, fragments Or.8212/480–484 found at the sites in the Lop Desert identified by Stein as L.A., L.C., L.E. and L.F., respectively; see the IDP Web site for a reconstruction.

39 London, British Library, Or.8212/168. "The fragments of 15 folios of the Salistambasutra were found before images around the Buddhist temple at Endere, deposited to propitiate the divinities. They must date from the Tibetan occupation of this part of Xinjiang and before the temple was abandoned in the mid-8th century" (V. Zwalf, *Buddhism: Art and Faith* [London: British Museum, 1985], no. 113).

40 See manuscripts from the Princeton University East Asian Library, Peald 7, and London, British Library, IOL Tib N 1061. See also Tsuguhito Takeuchi, "The Tibetan Military System and Its Activities from Khotan to Lop-Nor," in Whitfield, *The Silk Road* (note 4), 50–56, Fig. 2.

41 See the papers by J.-P. Drège and Akao Eikei in Susan Whitfield, ed., *Dunhuang Manuscript Forgeries* (London: British Library, 2002).

42 "Ein neuer Beitrag zur Geschichte des Papiers," in *Sitzungsberichte der kaislichen Akademie der Wissenschaften in Wien, philosophisch-historische Klasse* 148 (Vienna, 1904); and "Über die ältesten bis jetzt aufgefundenen Hadernpapiere," in *Sitzungsberichte der kaislichen Akademie der Wissenschaften in Wien, philosophisch-historische Klasse* 168 (Vienna, 1911).

43 The second edition of this work, published in 2004, summarizes the more recent archaeological discoveries of paper in China, suggesting that the invention of rag paper certainly dates back to the second century and possibly even the third cen-

tury BCE (145–50). T. T. Tsien discusses the spread of paper in his chapter, "Paper and Printing," in Joseph Needham, *Science and Civilisation in China*, 293–359 (Cambridge: Cambridge University Press, 1985 [rev. ed., 1987]).

44 "An Examination of Early Chinese Paper," *Restaurator* 4 (1979): 129–51.

45 "On the Origin of Papermaking in the Light of the Newest Archaeological Discoveries," *IPH Information* 15, no. 2 (1981): 38–47.

46 See note 36 and "Chronological Classifications of Dunhuang Buddhist Manuscripts" in Whitfield, *Dunhuang Manuscript Forgeries* (note 41). One of the few groups building on this work were the scholars Ueyama Daishun and Kudara Kogi, and more recently Mitani Mazumi. The latter presented a paper entitled "New Results of Research on the Chinese Buddhist Texts of the Berlin Turfan Collection: Plan of a New Chronological Standard of the Chinese Buddhist Manuscripts Excavated in Turfan" at a 2005 workshop in Berlin, "Digitalisierung chinesischer, tibetischer, syrischer und Sanskrit-Texte der Berliner Turfansammlung" (available at http://turfan.bbaw.de/bilder/workshop2005.pdf). For Kudari's work, see also Wolfgang Voigt, *Chinesische und Manjurische Handschriften und seltene Drucke: Teil 4* (Chinese Buddhist Texts from the Berlin Turfan Collections, Volume 3) (Stuttgart: Franz Steiner Verlag, 2005).

47 Among numerous studies, some cited above, see also Jean-Pierre Drège, "Notes codicologiques sur les manuscrits de Dunhuang et de Turfan," *Bulletin de l'École française d'Extrême-Orient* 74 (1985): 485–504.

48 See paper in Whitfield, *Dunhuang Manuscript Forgeries* (note 41).

CHAPTER 10

Providing Access to Manuscripts in the Digital Age

PETER M. SCHARF

As the principal medium of knowledge transmission shifts from handwritten and printed materials to the Internet, scholars increasingly expect to be able to find what they are looking for within seconds in an Internet search. In order to meet the expectations of today's Internet users while at the same time providing high-quality access to detailed information about Sanskrit manuscripts, the Sanskrit Library developed a pipeline to produce integrated hypertext access to manuscript metadata and images. Until now access to Sanskrit manuscripts has been severely hampered by distance to collections, isolation of artifacts from complementary research materials, deficiency or lack of metadata, and disarray within collections and within individual items of a collection. Arranging, cataloguing, scanning, and web-hosting of digital images of artifacts obviously address these problems. Yet information processing technology that developed primarily in the environment of the Roman alphabet and retains conventions that take uniform European linguistic representations for granted is still unable to handle features of non-European languages such as nonalphabetic scripts, multiple scripts, unusual orthographic conventions that hide word boundaries, and highly inflected and agglutinative language structures. As a result, the normal functionality of finding aids is inadequate to cope with Indian collections. The fact that manuscripts that are difficult to navigate must be used on site in special collection rooms isolated from related materials makes sought-for passages difficult and time-consuming to find. In order to address these issues, the Sanskrit Library has developed protocols, formats, and software to overcome the above-mentioned linguistic impediments and to provide web access to the primary cultural heritage materials of India. Materials developed include a comprehensive integrated hypertext catalogue and software to integrate digital images of manuscript pages with the corresponding machine-readable text, thereby providing direct and focused access to specifically sought passages on individual manuscript pages. The facilities for searching expected of contemporary web interfaces is thus extended to digital manuscript images, and the path is opened for generalized information extraction and search techniques to reach Sanskrit manuscripts. The Sanskrit Library's integrated hypertext access system may be implemented for any manuscript collection.

Crisis in the Transmission of Inherited Knowledge

The enormous heritage of knowledge and culture in perishable manuscripts written in Sanskrit in India is under threat of extinction as the dominant medium for the transmission of knowledge shifts and the life span of extant manuscripts approaches expiration. Just as the shift from handwriting to print drew practices and resources away from the culture of manuscripts, the current shift from the printed medium to the digital medium further marginalizes manuscripts as the expected methods of accessing information depart further from the norms of the manuscript culture. The government of India's National Mission for Manuscripts (www.namami.org) and forward-looking manuscript libraries around the world have recognized the importance of surveying, cataloguing, and making digital images of extant manuscripts as well as of encouraging critical text-editing. Yet the digital medium offers many facilities that could be engaged in order to enhance access to these valuable artifacts of India's heritage if technologies are adapted and extended to cope with the features of these items. A glance at the history of the preservation and loss of knowledge in previous media transitions and the adaptation of technologies needed to preserve knowledge during such transitions offers some insight into what is required to preserve the knowledge in manuscripts in general in the current transition to the digital medium. An investigation of such issues prompted the Sanskrit Library to undertake to adapt standards and develop formats, protocols, and innovative technologies to enhance access to Sanskrit manuscripts in the digital age. In this line, we adapted the Unicode standard, articulated phonetic encodings of Sanskrit, created transcoding software to and from various input and display methods, wrote linguistic software, and created a pipeline for dynamic cataloguing and text-image alignment in order to provide integrated digital access to Sanskrit manuscripts. The Sanskrit Library is eager to share its expertise in this area to help preserve the precious inherited knowledge and culture of India and to transmit it to future generations.

Sanskrit Literature

Sanskrit is the primary culture-bearing language of India, with a continuous production of literature in all fields of human endeavor over the course of four millennia. Preceded by a strong oral tradition of knowledge transmission, records of written Sanskrit remain in the form of inscriptions dating back to the first century BCE. In surveys to date, the National Mission for Manuscripts has already counted more than five million manuscripts, and David Pingree, the renowned manuscriptologist and historian of mathematics, estimated that extant manuscripts in Sanskrit number over thirty million—more than one hundred times those in Greek and Latin combined—constituting the largest cultural heritage that any civilization produced before the invention of the printing press. Sanskrit works include extensive epics; subtle and intricate philosophical, mathematical, and scientific treatises; and imaginative and rich literary, poetic, and dramatic texts. While the Sanskrit language is of preeminent importance to the intellectual and cultural heritage of India, the importance of the intellectual and cultural heritage of India to the rest of the world during the past few millennia and in the present era can hardly be overestimated. Indian culture has been a major factor in the development of the world's religions, languages, literature, arts, sciences, and history. The tradition of Vedic recitation, dating to the second millennium BCE, was declared by the United Nations Educational, Scientific and Cultural Organization (UNESCO) in November 2003 to be one of the "masterpieces of the oral and intangible heritage of humanity" under a program aiming to raise public awareness of the value of this heritage and encourage governments to take legal and administrative steps to safeguard it. In the first millennium BCE trade flourished between India and the Achaemenid Empire, Hellenistic empires, and the Roman Empire. In the early centuries of the Common Era, political, educational, and religious leaders brought Indian literature and culture to Southeast Asia. Buddhist missionaries brought Indian culture to Tibet, Central Asia, and China, and from there to Korea and Japan. Through the intermediary of Muslim scholarship, and Latin and Greek translations of it in the eleventh, twelfth, and thirteenth centuries, Indian astronomy, astrology, mathematics, medicine, philosophy, and literature served as the sources of the revival of civilization in the Latin West and in Byzantium. Indian ideas permeate the scientific texts of the high Middle Ages from which modern Western science and literature directly descend.

The Crisis

As the life span of manuscripts in India elapses, the skills required to use them become rarer, and the attention of those interested in their contents becomes restricted to materials available on the Internet, it has become critical to provide digital access to manuscript contents. Manuscripts in India are kept in a variety of conditions ranging from climate-controlled libraries that are part of university campuses, government institutions, or esteemed societies, to temple libraries, private libraries, and small caches in private homes. In many of the latter types of repositories, vulnerable manuscripts are exposed to a range of temperatures and high humidity and are unprotected from insects and worms. The physical support of manuscripts in India is generally palm leaf in the south and paper in the north. These materials last for three hundred to five hundred years. More than half the life span of most manuscripts has already expired because the tradition of copying manuscripts by hand diminished steadily after the introduction of moveable type in India in the last quarter of the eighteenth century and has now all but ceased.

The number of traditionally trained pandits and modern scholars of Sanskrit is diminishing both within and outside India with the lapse of time and changing educational trends. In India, the status of Sanskrit in most school systems has receded to that of a foreign language. Abroad, primary and secondary education remains grossly undersupplied with adequate educational materials about India so that few students are aware of its rich and abundant literature. Manuscript materials are completely inaccessible to students at these lower levels. As a result of low awareness, few colleges and universities train students in the languages of India. Even at centers of Indological research, popular trends in the humanities and social sciences often assume precedence over philology, paleography, and manuscriptology. These educational trends are spreading to India as well. Neglect of manuscripts by scholars contributes to neglect of manuscripts by their custodians, resulting in peril to these valuable and unique artifacts of the heritage of India. Because the process of making critical editions is demanding and time-consuming, the few scholars engaged in the process cannot possibly exhaust the work of collating all the extant manuscripts in critical editions within the remaining manuscript life span. Therefore, the knowledge in manuscripts is in danger of perishing with its aging paper and palm-leaf substrates. Action must be taken to preserve this valuable knowledge and cultural heritage and to ensure its accessibility in the new dominant medium of knowledge transmission.

Media Transitions

The current transition of the dominant medium of knowledge transmission from printed book to electronic text is not the first transition in the medium of knowledge. History records two other such transitions: the transition from oral tradition to writing, and the transition from manuscript to printed text. The transition from oral tradition to writing is recorded in ancient Greece by Plato in the fifth century BCE. In a passage in Plato's *Phaedrus* (275a), for example, Socrates disparages writing by relating the words of King Thamus of the Egyptian Thebes to the god Theuth when Theuth revealed the art of writing to him. When Theuth promised that it would make the people wiser and improve their memories, King Thamus retorted that it would have the very opposite effect: "It will implant forgetfulness in their souls; they will cease to exercise memory because they rely on that which is written." While there are both benefits and detriments to the medium of writing vis-à-vis oral transmission, this passage of Plato's recognizes the introduction of writing into disciplines of learning in Greece and recognizes Egyptian influence in this introduction.

Writing was introduced earlier, at the end of the fourth millennium BCE in Sumeria and Egypt. The earliest documents record economic transactions, such as the number of sheep sold or the numbers of bundles of grain collected in taxes. In India, while the Harrapan script remains undeciphered, the earliest extant uses of Kharoṣṭhī and Brāhmī scripts are on public monuments and in edicts. Aśoka commissioned the Brāhmī script for these administrative purposes during the expansion of his empire in the latter half of the third century BCE.

The introduction of writing was originally for administrative and economic purposes; it was not used initially for literary or scientific affairs. Only later did literary composition make the transition into the written medium. In India, the earliest inscriptions are in Prākrit, not in Sanskrit. The earliest Sanskrit inscriptions date to the first century BCE— two centuries later than the oldest inscription in Prākrit. The Vedic tradition and the core sūtra texts in Sanskrit continued to be transmitted orally for millennia after Aśoka introduced writing for edicts in Prākrit, even as writing on paper and palm leaf became the principal means to distribute knowledge. Public performance also continued to be widely popular even as writing spread. Oral learning diminished gradually in educational systems around the world up to the present day. In India, oral learning remains alive only in traditional Vedic educational institutions (*pāṭhaśālās*). In the West, it has ceased even in language instruction and remains essential only in classes in the dramatic arts. Writing gradually crept from administrative into literary uses and overtook orality as the dominant mode of transmission in education.

Moveable type was invented by Gutenberg in 1445. While the first typeface was used in a Latin textbook, it was very soon adopted for literary purposes. The Gutenberg Bible was printed just ten years later in 1455. Types were first employed in India to print Christian doctrine at a Jesuit printing press in Goa, which operated between 1556 and 1674. Tamil types were created there in 1578. Devanāgarī types were first created in Rome in 1771 to print the *Credo* in Hindi, and Charles Wilkins created a Devanāgarī typeface soon afterward that was used to print his *A Grammar of the Sanskrita Language* in 1808. A Bengali grammar was printed in Hoogly in 1778. Printing gradually replaced manuscript copying as the dominant mode of knowledge transmission in India during the nineteenth and twentieth centuries. The first examples of printed typeface tended to imitate handwritten characters, and printers' type cases, which counted hundreds of characters, included types for numerous ligatures. Characters were standardized and repertoires reduced to accommodate the restrictions of new technologies such as hot-metal typesetting and the typewriter.[1]

Preservation and Loss of Knowledge in Media Transitions

Knowledge exists fundamentally in the consciousness of knowledgeable people. They express and communicate that knowledge through speech, visual images, and performance and often combine these various means of communication. Oral communication is often accompanied by gestures, for instance. Each of these modes of communication has the potential to imitate the others. When a mode of communication is copied into another medium, the copy can be no better than its original. The copy selectively reproduces what the medium of reproduction permits and the copyist chooses to include. The transcription of a lecture will not include the gestures of the speaker nor his changes in intonation. In the same way, a manuscript reproduction of oral recitation of a Vedic text will lose much. Head and hand gestures and voice fluctuations will not be recorded. Yet special effort on the part of the copyist may extend the target medium to accommodate unusual information. The character set of ordinary written Sanskrit, for instance, was extended by marks used to capture pitch variation in Vedic. Yet if subsequent readers of the copy fail to understand the significance of certain marks, they will cease to understand what those marks represent in the original mode of communication. If no one remains to explain their significance, that information will be lost. The significance of Vedic accent marks in less common traditions, such as in the Kāṭhaka and Maitrāyaṇī branches of Yajurveda, and Rāṇāyanīya and Jaiminīya branches of Sāmaveda, is known to few and requires some research to discover. Moreover, transcription of these Vedic texts into Roman script often obliterates the differences between different traditions of recitation.[2] The knowledge of these accentual traditions would perish with the death of a few individuals and the loss of a few volumes.

In general, knowledge gets lost in media transitions because a newer medium cannot accommodate all the information present in the older medium it copies; people fail to encode information accurately in the new medium; younger generations, accustomed to the new medium, cease to learn to access information in the old medium; and the substrate of the old medium perishes. Preserving knowledge during media transitions requires special attention to counter each of these points. It requires copying the most original form of the information, adapting the new medium to accommodate the desired information, creating methods to accurately encode desired information, adapting the presentation of old information to meet new standards of access, and acting in a timely fashion before the old medium perishes. Preserving knowledge in media transitions requires recognizing that the new medium is not a static inheritance. Intelligence, creativity, and effort can adapt the new medium to meet the needs of the information that is desired to be expressed.

New media provide technological advances that offer new possibilities for the propagation of knowledge. Writing endures for a considerable length of time, while speech vanishes the moment after it is uttered. Printing allows the wide distribution of multiple exact copies with relatively little effort. Alphabetization is a technology appropriate to visual media such as writing and print. Merely by the shared standard of the alphabetic order, one is able to locate sought items in a dictionary or index. Thesauri, typically memorized in oral medium, required much greater effort to learn, yet once learned, they allowed instant random access to their contents; without the oral medium and memorization, they require much greater effort to access. Digital technology, however, delegates alphabetization to software and replaces manual use of alphabetized dictionaries and indices with the search interface. A digital search interface allows random access without memorization. Recognizing the potential of the new medium and utilizing it to its fullest are essential for the preservation and propagation of knowledge. At the least, users' expectations of the new medium of knowledge transmission need to be met lest information inaccessible by the methods to which the users are accustomed simply gets disregarded, neglected, and lost.

Expectations Regarding Information Access in the Digital Medium

Digital technology, computational linguistic methods, and the Internet allow easier and faster access to information. Digital technology allows the magnification of text and images and enhancement of images through the adjustment of lighting and contrast. HTML interfaces permit greater synthesis and integration of information via linking than does linear text in the printed or written medium. The digital medium also allows information to be represented in various views without significant additional labor or expense. The digital medium allows easy access to greater detail and to obscure sources where access to physical copies would require prohibitive expense and effort.

Internet users expect to find what they are looking for on the Web within seconds. Yet the seamless fulfillment of their expectations depends upon information-processing technology that has developed primarily in the environment of the Roman alphabet. Functionality that is taken for granted for European languages has not yet been developed for other languages. One expects to be able to search a PDF file. One expects to be able to run optical character recognition software on a PDF file to extract machine-readable text. One expects to find material so extracted from PDF files in a general Web search interface. Generalized information extraction and search techniques cannot adequately handle literary materials for which there is a lack of adequate optical character recognition software, inconsistent encoding, obscure word boundaries, complex morphology, and free syntax. However, the Sanskrit Library is developing just such tools for the principal culture-bearing language of India.

Overcoming Obstacles to Access of Indian Heritage

The Sanskrit Library has and continues to develop the techniques and infrastructure necessary to integrate Sanskrit manuscripts embodying primary cultural heritage materials of India with digital text, lexical resources, and linguistic software. In a project funded by the U.S. National Science Foundation, the Sanskrit Library standardized Sanskrit text encoding, revised the Unicode Standard to include characters necessary for Indic cultural heritage, supplied validated data for optical character recognition, prepared the major digital Sanskrit-English lexicon for integration with linguistic software, produced several other digital lexical resources, produced a full-form Sanskrit lexicon and morphological analyzer, and fostered international collaboration in the area of Sanskrit computational linguistics.

In *Linguistic Issues in Encoding Sanskrit*, Malcolm Hyman and I completed a comprehensive survey of linguistic and theoretical issues related to the encoding of Sanskrit language and designed accurate, principled phonetic encoding schemes for Sanskrit linguistic processing. Although Indic scripts reflect the phonetics of Sanskrit transparently, the orthography of the various semisyllabic Brāhmī-based scripts of India departs from an ideal one-to-one coding of Sanskrit sounds. Yet the sophisticated linguistic traditions of India provide direct access to the phonology of the language, thereby allowing the creation of encodings ideal for linguistic purposes. We designed three phonetic encodings based upon different principles: Sanskrit Library Phonetic basic (SLP1), which uses only ASCII codes; Sanskrit Library segmental (SLP2), which has a unique code point for each phonetic segment, regardless of accent, nasalization, length, or other feature; and Sanskrit Library featural (SLP3), which encodes only the features that characterize sounds rather than phonetic segments.

While the encoding schemes we designed are suited to linguistic processing, users have their own preferences for reading and keyboard entry. To interface with the schemes preferred by users, the Sanskrit Library developed transcoding routines to translate between its phonetic encodings and standard and popular encodings used for data entry and display. A preferences menu permits users to select desired input methods, such as Kyoto-Harvard, ITrans, Velthaus, and WX, and to display content in a variety of scripts, including the major scripts of India, and standard Romanization.

After an investigation of Sanskrit paleography, I initiated worldwide collaboration to extend the Unicode Standard to include characters required for the proper display of the ancient Vedic heritage texts of India. Partners included the Indian Ministry of Communications and Information Technology, the Centre for Development of Advanced Computing in Mumbai, and the Script-Encoding Initiative at Berkeley. As a result of these efforts, Unicode Standard version 5.2 incorporated sixty-eight additional characters in two code blocks, Devanagari Extended and Vedic Extensions, both accessible under South Asian Scripts via the Unicode Character Code Charts page (www.unicode.org/charts).

By running our inflection software on the 170,000 nominal and verbal headwords in Monier-Williams's *A Sanskrit-English Dictionary* (Oxford: Clarendon, 1899), the most complete English-language dictionary of Sanskrit, we created a full-form lexicon of eleven million entries that associates each inflected form with its inflectional identifier and headword. The full-form lexicon allowed us to build a morphological analyzer. The analyzer displays all possible analyses of the inflected nominal form entered in the analyzer input field. Each analysis consists of the inflectional identifier and stem, the latter of which is a link to the Sanskrit Library multidictionary interface.

Another project jointly funded by the U.S. National Endowment for the Humanities (NEH) and the Deutsche Forschungsgemeinschaft extended the Sanskrit Library's multidictionary interface by integrating supplements to the major bilingual dictionaries already included and by adding specialized dictionaries, indigenous Indian monolingual dictionaries, and traditional linguistic analyses. We hope to add traditional thesauri and additional analyses in a sequel.

The Sanskrit Library obtained three hundred digital editions of texts from the Thesaurus Indogermanischer Text- und Sprachmaterialien at Johann Wolfgang Goethe Universität in Frankfurt am Main (titus.uni-frankfurt.de), the Vedic Reserve at Maharishi University of Management (is1.mum.edu/vedicreserve), the NEH-funded grammatical data-bank project headed by George Cardona at the University of Pennsylvania in the early 1990s, and other sources, and displays them in a reader page. Each word in texts in which interword phonetic changes (*sandhi*) have been analyzed dynamically links to the morphological analysis window, where stems link to the multidictionary interface. This integration of digital texts, linguistic software, and lexical sources thus provides an environment in which users can easily access resources to assist in studying the texts they wish to read.

The Sanskrit Library texts additionally are integrated with Gérard Huet's Sanskrit Heritage Site parser using distributed interoperable Web services. In texts in which sandhi has not been analyzed, each sentence is a link to the Sanskrit Heritage Site's parser (sanskrit.inria.fr/DICO/reader.en.html). The Sanskrit Heritage Site additionally allows one to submit compounds for further analysis by the compound analyzer built by Amba Kulkarni at the University of Hyderabad and to submit analyzed sentences for syntactic analysis by her dependency tree parser. Encouraged by the success of this sort of distributed international cooperation, Huet, Kulkarni, and I collaborated with colleagues in forming the Sanskrit Computational Linguistics Consortium to hold symposia, workshops, and seminars in order to foster collaborative research and resource sharing in Sanskrit natural language processing. Five symposia were held from 2007 to 2013 in the United States, France, and India, with papers published by Springer and D. K. Printworld.[3]

In an NEH-funded project, that took place between 2009 and 2012 entitled "Enhancing Access to Primary Cultural Heritage Materials of India: Integrating Images of Literary Sources with Digital Texts, Lexical

Resources, Linguistic Software, and the Web," the Sanskrit Library aimed to enhance access to primary cultural-heritage materials of India housed in American libraries by integrating them with the digital texts, lexical resources, and linguistic software in the Sanskrit Library. The project developed protocols, formats, and software to integrate into its digital library digital images of 160 Sanskrit manuscripts, numbering 25,000 pages, that represent two central Indic texts, *Mahābhārata* and *Bhāgavatapurāṇa*, in the Brown University and University of Pennsylvania Libraries. The project developed a comprehensive dynamic catalogue that allows access to manuscripts via numerous criteria and explored text-image alignment techniques to permit focused access to particular passages on manuscript pages by way of searching for the passage in corresponding digital text.

Dynamic Cataloguing

Library Catalogues

A typical library catalogue includes a bare minimum of information regarding manuscripts. For example, the Penn Libraries' online catalogue, Franklin, includes only the following categories: the title, the collection to which the manuscript belongs, a reference number, and a description consisting of the extent, material, dimensions, language, and script.

American Committee for South Asian Manuscripts

The American Committee for South Asian Manuscripts (ACSAM) adopted much more thorough cataloguing standards. ACSAM was established by the late David Pingree in 1995 under the aegis of the American Oriental Society for the purpose of preserving and promoting access to the manuscripts of South Asia held in North American collections. ACSAM prescribed the collection of data in twenty-three categories for the purpose of creating complete descriptive catalogue entries. These categories include explicit provision for bibliographic information of editions of the work and catalogues that mention the work; information about scribes, patrons, and owners; description of the condition of the manuscript, its binding, format, and illustrations; additions to it; commentaries it may contain; the transcription of the beginning and end of the text; the closing (such as *iti śrīmadbhagavadgītā samāptā*, called there "colophon"); and scribal trailers (called there "post-colophon").

The Sanskrit Library's TEI-Ms Template

The "Enhancing Access to Primary Cultural Heritage Materials of India" project developed an XML manuscript-cataloguing template in accordance with the guidelines of the Text Encoding Initiative's (TEI) Manuscript Description (www.tei-c.org/release/doc/tei-p5-doc/en/html/MS.html). The XML template incorporates the comprehensive standards of manuscript description and classification set by ACSAM in a standard markup that allows digital processing and facilitates Internet access. The categories included in the Sanskrit Library's manuscript-cataloguing template are described in detail in Appendix A. These categories include subject classification in accordance with the Library of Congress subject headings, a subject classification in accordance with a traditional Indian knowledge map, and a classification of meters. The *encodingDesc* element defines the term of reference to the Library of Congress subject headings and provides a definition of symbols used in the standard classification of Sanskrit metrical patterns. The *profileDesc* element includes elements in which to list Library of Congress subject headings that describe the manuscript and to classify the manuscript in accordance with the Sanskrit Library's own Indic subject classification. The latter, shown in Appendix B, is based upon well-known traditional divisions of disciplines. The manuscripts of the *Bhāgavatapurāṇa*, for instance, are classified as Purāṇa. The classification of the manuscripts of parts of the *Mahābhārata* subsumes them under the category Itihāsa. Crucially, the

Sanskrit Library catalogue provides transcription of identifying passages of a manuscript that permit works of interest to be located even where variations in, or absence of, the title and author's name would render the work unlocatable in an ordinary library catalogue.

Transcription

The Sanskrit Library TEI manuscript catalogue entry files include identifying passages, such as rubrics, incipits, explicits, final rubrics, and colophons, by referring to transcriptions of these passages in the *text* element in the body of the TEI document. Elements are used there to indicate the following: text divisions (*div*), headings (*head*) and closings (*trailer*), speeches (*sp*), the introduction of speakers (*speaker*), verses (*lg*) and lines of verse (*l*), paragraphs (*p*), sentences (*s*), other segments of text such as verse quarters or sections of paragraphs or trailers (*seg*), gaps in the manuscript that may result from damage or missing leaves (*gap*), and bibliography corresponding to transcribed passages (*bibl*). Some of these text markup elements are supplied with XML identifier attributes (xml:id) that permit them to be the target of reference. Elements that describe the textual contents of each work in the "*msContents*" element of the TEI header (*rubric*, *incipit*, *explicit*, finalRubric, and colophon) make formal reference to the relevant textual elements by the use of the "*corresp*" attribute. These passages can then be indexed and made searchable in an HTML catalogue interface. Bibliographic references from identified passages in published works may include *corresp* attributes that refer to the passage in a digital edition of the text. An HTML interface can then link to the relevant passages directly.

The extensive markup used in the Sanskrit Library manuscript catalogue template allows catalogue entries to be indexed by numerous criteria in order to provide an extremely versatile catalogue index interface. It is not necessary, however, that all these details be provided for every manuscript in order to include the entry in the catalogue and to produce a handsome display of the catalogue entry. The Sanskrit Library developed software to format whatever information is provided automatically in an HTML display and to include that information automatically in a dynamic catalogue index. The latter will be described shortly after a word about how the catalogue entry is linked to manuscript images.

Facsimiles

Although the manuscript catalogue just described provides a great deal of information to scholars, one of the principal functions of a catalogue is to provide access to manuscripts themselves, for it is the manuscripts that are the ultimate witnesses to original written works. In addition to providing comprehensive and versatile access to catalogue entries, the manuscript project provides focused access to passages of interest in manuscript images. The project produced high-quality images of the manuscripts catalogued. In palm-leaf manuscripts, each palm leaf was imaged individually with a color-balancing measuring stick. In unbound paper manuscripts where leaves are often attached in pairs, pages were imaged in pairs, the verso of the first folio with the recto of the next. In order to allow more focused Web display, paired JPEG images were split and all images automatically cropped to within a few millimeters of their edges using software developed by Donglai Wei for the project.

The alignment of manuscript pages with photographic images requires careful control if a scholar is to access the sought images in a rational manner. While manuscript photographers assign names in sequence, such as UPenn490_0001, UPenn490_0002, and so on, descriptive names refer to the folio and side, such as "f. 1r" for the recto of the first folio. Anomalies occur in the association of a directory of the images of a manuscript with the enumeration of leaves in the foliation element when, for example, the directory of images includes images of accompanying documents and bindings that are not included in the *foliation* element. Anomalies also occur when duplicate images are delivered or when pages are inadvertently skipped.

Besides splitting the images, Wei's software produced an XML file containing references to the images and their cropped zones in accordance with the TEI guidelines. Ralph Bunker, the technical director of the Sanskrit Library and software engineer for the NEH project, developed software to assign descriptive names to each page image in conformity with the *foliation* element in the *teiHeader* manuscript description. Bunker developed the Folio software to compare image references with directory contents and generate a report of misalignments. Folio then produces an HTML display that allows human validation of image references by comparison with miniature images and with page references in annotations produced by the SITA software described below (see Figure 10.11). Cataloguers adjust misaligned names by editing an XML page directory to indicate anomalies. Once references are validated, Bunker's catalogue preparation software inserts the TEI graphic references with their descriptive name identifiers into the *facsimiles* element in the TEI manuscript catalogue entry file. The images become available to scholars via links from the descriptive names in HTML pages produced from TEI entry files by software described in the next section.

The Sanskrit Library Manuscript Catalogue HTML Display

The Sanskrit Library Siva software generates HTML pages from the TEI XML catalogue entry files. The software neatly formats the catalogue data, graphic references, and text in the XML file and automatically generates hyperlinks. Figure 10.1 shows the HTML display of the contents of the manuscript UPenn 490 containing the *Bhīṣmastavarāja*. The HTML display is produced programmatically from the TEI XML source file, the first half of the *msItemStruct* element of which is shown in Figure 10.2. Links in the HTML display lead from notes in the catalogue description to transcribed passages, from XML graphic references in the *facsimiles* section to digital images, from title abbreviations to full bibliographic descriptions, and from bibliographic scope references to corresponding digital texts and annotated manuscript images. For example, clicking on the verse number *127* in the beginning of the note in Figure 10.1 scrolls to the *transcription*

Contents

Work 1 (complete)

Locus:	ff. 1r–17v
Author:	Kṛṣṇa Dvaipāyana Vyāsa
Title:	**Bhīṣmastavarāja**
Incipit:	f. 1v .. janamejaya uvāca .. .. śaratalpe śayānas tu bhāratānāṁ pitāmahaḥ katham utsṛṣṭavān dehaṁ kaṁ cid yogam adhārayat ..1.. (Anuṣṭubh) (MBh. 12.47.1)
Explicit:	f. 17r stavarājaḥ samāpto yaṁ viṣṇor adbhutakarmaṇaḥ gāṁgeyena purā gīto mahāpātakanāśanaḥ ..127.. (MBh. 12.47.65*98, lines 3–4) (Anuṣṭubh)
Final rubric:	f. 17r–f. 17v iti śrīmahābhārate satasahasryāṁ saṁhitāyāṁ śāṁtiparvaṇi rājadharme bhīṣmoktaṁ stavarājastotraṁ saṁpūrṇam ..
Colophon:	none
Note:	The concluding verse 127 on f. 17r, 'Finished is this sin-destroying regal praise of Viṣṇu of wonderful deeds previously sung by the son of the Ganges (Bhīṣma)', and the final rubric on f. 17v, 'Finished is the regal hymn of praise uttered by Bhīṣma in the *Rājadharma* section in the *Śāntiparvan* in the *Mahābhārata* of a hundred thousand verses', identify the text as the *Bhīṣmastavarāja*, which comprises the contents of Adhyāya 47 in the *Rājadharma* section of the *Śāntiparvan* of the *Mahbharata* (MBh. 12.47.1–12.47.72). Bhīṣma sings this praise of Viṣṇu just prior to abandoning his arrow-filled body. The manuscript is 75% longer than the text in the Pune critical edition. The last two verses of the text (MBh. 12.47.71–72) in the critical edition correspond with verses 119–120 on f. 16r of the ms., though the latter varies considerably. While the text of the chapter in the critical edition ends there, the ms. continues with six additional verses, four of which correspond with verses in the star passages in the critical apparatus: verse 121 on f. 16r corresponds with the first two lines of star passage 98 inserted by M1.3 after MBh. 12.47.65 while the concluding verse 127 on f. 17r corresponds with lines three and four of star passage 98. Verse 122 on f. 16v corresponds with lines 5–8 of star passage 94 inserted by K4 and Dn after MBh. 12.47.60, verse 124 on f. 16v corresponds with lines 5–8 of star passage 99 inserted by D7, T, and G1.2 after MBh. 12.47.72, and verse 125 on f. 16v–f. 17r corresponds with lines 1–4 of the same passage. Verses 123 (in Anuṣṭubh meter) and 126 (in Mālinī meter) are not found in the critical text or apparatus.
Language:	Sanskrit in Devanāgarī script

```
<msContents>¬
 <msItemStruct defective='false'>¬
  <locus from='1r' to='17v' target='#images'>ff. 1r--17v</locus>¬
  <author xml:lang='sa-Latn-x-SLP1'>{k}fzRa {d}vEpAyana {v}yAsa</author>¬
  <title type='main' xml:lang='sa-Latn-x-SLP1'
      rend='bold'>{B}IzmastavarAja</title>¬
  <title type='context' xml:lang='sa-Latn-x-SLP1'
      rend='bold'>{m}ahABArata</title>¬
  <title type='context' xml:lang='sa-Latn-x-SLP1'
      rend='bold'>{S}Antiparvan</title>¬
  <incipit xml:lang='sa-Latn-x-SLP1' corresp='#sp1'>¬
   <locus from='1v' facs='#f1v' xml:lang='en'>f. 1v</locus>¬
  </incipit>¬
  <explicit xml:lang='sa-Latn-x-SLP1' corresp='#v127'>¬
   <locus from='17r' facs='#f17r' xml:lang='en'>f. 17r</locus>¬
  </explicit>¬
  <finalRubric xml:lang='sa-Latn-x-SLP1' corresp='#fr'>¬
   <locus from='17r' to='17v' facs='#f17r #f17v' xml:lang='en'>f. 17r--f.
       17v</locus>¬
  </finalRubric>¬
  <colophon>none</colophon>¬
  <note xml:id='BhSRdesc'>The concluding verse <ref
      target='#v127'>127</ref> on <locus from='17r' facs='#f17r'>f.
      17r</locus>, <q>Finished is this sin-destroying regal praise of
      <persName type='character'
      xml:lang='sa-Latn-x-SLP1'>{v}izRu</persName> of wonderful deeds
      previously sung by the son of the Ganges (<persName type='character'
      xml:lang='sa-Latn-x-SLP1'>{B}Izma</persName>)</q>, and the <ref
      target='#fr'>final rubric</ref> on <locus from='17v' facs='#f17v'>f.
      17v</locus>, <q>Finished is the regal hymn of praise uttered by
      <persName type='character'
      xml:lang='sa-Latn-x-SLP1'>{B}Izma</persName> in the <title
      type='mentioned' xml:lang='sa-Latn-x-SLP1'
      rend='italic'>{r}AjaDarma</title> section in the <title
      type='mentioned' xml:lang='sa-Latn-x-SLP1'
      rend='italic'>{S}Antiparvan</title> in the <title type='mentioned'
```

section to show verse 127 of the *Bhīṣmastavarāja* in context. Clicking the manuscript page reference *fol. 17r* in the same note opens an image of the recto of folio 17. The link to this image is also found listed in order in the *facsimiles* section, and in the manuscript view of the *transcription* section. Clicking the title abbreviation *MBh.* in the bibliographic reference at the end of the incipit in the content section shown in Figure 10.1 displays a bibliographic description of the Pune critical edition of the *Mahābhārata*. Finally, clicking the bibliographic scope reference *12.47.1* (book [*parvan*], chapter [*adhyāya*], and verse numbers) at the end of the bibliographic reference in the incipit opens the alignment display interface shown in Figure 10.12. This figure displays the text of the Sanskrit Library's digital edition of the *Śāntiparvan* of the *Mahābhārata*, which contains the *Bhīṣmastavarāja*, scrolled to the beginning of the *Bhīṣmastavarāja*. The alignment display interface is described further below. (See Text-Image Alignment.)

Text in the Sanskrit Library manuscript catalogue HTML display is displayed in the *transcription* section in two layouts. Initially, the text is displayed in the text layout in accordance with the structure of the text as shown in Figure 10.3. Verses are formatted in metrical lines and verse quarters (*pādas*). Each sentence of prose is put on a separate line, as are headings and trailers, and bibliographic references are shown. Editorial corrections, notes, and references are displayed. Clicking the "Show manuscript layout" button displays the transcription in accordance with the page and line structure of the manuscript itself as shown in Figure 10.4. Each line in the manuscript is displayed on its own line, and the lines on a page are displayed beneath the folio number of that page. References, notes, and corrections are removed to avoid interrupting the original layout.

The folio page reference in the manuscript layout is a link to the image of that page. For instance, clicking *f. 1v* in Figure 10.4 opens an image

---

Figure 10.1. A partial view of the contents section in the Sanskrit Library manuscript catalogue HTML file for UPenn 490.

Figure 10.2. A partial view of the msContents element in the Sanskrit Library manuscript catalogue XML file for UPenn 490.

Show manuscript layout

Transcription (text structure)

śrīgaṇeśāya namaḥ ..
.. janamejaya uvāca .. ..
śaratalpe śayānas tu bhāratānāṁ pitāmahaḥ
katham utsṛṣṭavān dehaṁ kaṁ cid yogam adhārayat ..1.. (Anuṣṭubh) (MBh. 12.47.1)
vaiśaṁpāyana uvāca ..
śṛṇuṣvāvahito rājan śucir bhūtvā samāhitaḥ
bhīṣmas tu kuruśārdūla dehotsargaṁ samāśrayat 2 (MBh. 12.47.2) (Anuṣṭubh)
<..>
<..>
<..> (Anuṣṭubh)
śrībhīṣma uvāca ..
ārādhayāmy ahaṁ viṣṇuṁ vācāṁ jigadiṣāmy ahaṁ
tayā vyāsasamāsīnyā prīyatāṁ puruṣottamaḥ 13 (MBh. 12.47.10) (Anuṣṭubh)

Show text structure

Transcription (manuscript layout)

f.1v
Line 1: śrīgaṇeśāya namaḥ .. .. janamejaya uvāca .. .. śarata
Line 2: lpe śayānas tu bhāratānāṁ pitāmahaḥ katham utsṛṣṭavān de
Line 3: haṁ kaṁ cid yogam adhārayat ..1.. vaiśaṁpāyana uvāca .. śṛṇu
Line 4: ṣvāvahito rājan śucir bhūtvā samāhitaḥ bhīṣmas tu ku
Line 5: ruśārdūla dehotsargaṁ samāśrayat 2 <..> <..>

f.3r
Line 1: <..> śrībhīṣma uvāca ..
Line 2: ārādhayāmy ahaṁ viṣṇuṁ vācāṁ jigadiṣāmy ahaṁ tayā
Line 3: vyāsasamāsīnyā prīyatāṁ puruṣottamaḥ 13 <..>

f.15r
Line 2: <..> iti vidyātapoyonir ayonir viṣṇur ī
Line 3: ḍitaḥ vāgyajñenārcito devaḥ prīyatāṁ me janārddanaḥ ..11..
Line 4: vaiśaṁpāyana uvāca .. etāvad uktvā vacanaṁ bhīṣmas tadgata
Line 5: mānasaḥ namaskṛtyaiva kṛṣṇāya prayāṇam akarot tadā 12 <..>

Figure 10.3. A partial view of the text structure layout in the Sanskrit Library manuscript catalogue HTML file for UPenn 490.

Figure 10.4. A partial view of the manuscript layout in the Sanskrit Library manuscript catalogue HTML file for UPenn 490.

of the verso of folio 1. As in bibliographic references in the *content* section described above, in the text structure layout too, the abbreviation of the title (*MBh.* after verse 1 in Figure 10.3) is a link to the full bibliographic description, and the bibliographic scope reference links to the corresponding digital text displayed in the alignment display interface. For example, clicking *12.47.1* at the end of the bibliographic reference in Figure 10.3 opens the alignment display interface shown in Figure 10.12.

The Sanskrit Library Manuscript Catalogue Index

The Sanskrit Library manuscript catalogue index provides numerous avenues to locate manuscripts. These avenues are not limited to the ordinary criteria of author, title, institution, manuscript identifiers, language, script, and subject heading found in most library and manuscript catalogues, although these categories are also available in the initially displayed general pane of the index. The numerous categorical, keyword, and text searches the index provides go much further than just allowing one to locate known texts and manuscripts. They also provide a research tool to gather information about various aspects of manuscript culture ranging from the content of works and historical information to scribal practices and material culture.

The *identifiers* pane permits not just a detailed search by catalogues that mention the manuscript and by the housing collection and its identifier, but includes in addition menus to search by the city ("settlement"), institution, and repository separately in case the precise identifiers are not known.

The *person* and *title* panes include menus to search by various categories besides the usual author and title, the latter of which finds whatever particular title the housing library chose to give the manuscript. Hence, one can search by the name of an author or editor of a work mentioned in the catalogue description who is not necessarily the author or editor of the work included in the manuscript. One can search by characters that participate in a work, thereby allowing one to locate works regarding a theme of interest regardless of the title of the work. For example, one can search for Bhīṣma or Yudhiṣṭhira to locate works in which these characters participate. Or one can search for Hāhā or Hūhū to discover that the *Gajendramokṣaṇa* narrates the tale of how these two vying celestial musicians were cursed by the competition judge, whose judgment was doubted, to be born as a crocodile and an elephant. The ability to search for the names of scribes, owners, or any other persons permits one to research the prosopography of a person mentioned in a different context. The ability to search for date of origin and place of origin or provenance in the history pane offers similar facility.

The *content* pane (shown in Figure 10.5) allows one to search for a particular passage that occurs in a rubric, incipit, explicit, final rubric, colophon, or addition. One can also search for the abbreviated titles that accompany enumeration (signatures). One can search for manuscripts in the description of which certain Sanskrit terms have been used or in which the meter of transcribed verses has been identified. These search avenues may assist scholars in identifying the content of other manuscripts as well as in finding manuscripts in the Sanskrit Library. For example, if one is

Figure 10.5. The *content* pane in the Sanskrit Library manuscript catalogue index.

attempting to identify the work in a manuscript for which no title or final rubric is available but which has *nā* written above the number on the verso of each folio, one can select *nā* from the *signature* menu on the *content* pane in the Sanskrit Library manuscript catalogue index, click the *submit* button to find that UPenn 2639 contains such a signature, click the link *penn2639.html* listed below to open its HTML catalogue entry, scroll down to the *physical description* section, note under the *signatures* label that *nā* is similarly used in work 5 of this manuscript, scroll up to the *content* section and discover in the description of work 5 that it is titled *Nāgalīlā*. If one were unable to identify the work in a manuscript missing its first and last pages whose second folio began *mayā hi deva*, one could search for the passage in the incipit's text box on the *content* pane of the Sanskrit Library manuscript catalogue index and quickly locate five manuscripts of the *Gajendramokṣaṇa* that begin with these words.

The *layout*, *hand*, and *decoration* pane allows one to search for manuscripts by the number of lines, colors, or certain keywords used to describe different hands and border decorations, as well as to search for illustrations. One can select, for instance, the illustration description that begins, "The ms. contains five full-color illustrations," to locate UPenn Ms. Indic 5 (Figure 10.6), click the link *msindic5.html* to open its HTML catalogue entry, scroll down to the *decoration* section, read the descriptions of the illustrations there (Figure 10.7), and click on the folio link *f. 206v* to open the manuscript page accompanying the narration of the *Gajendramokṣaṇa* that shows the illustration of Viṣṇu accompanied by Garuḍa liberating the elephant from the crocodile in the lake (Figure 10.8).

The *physical* pane allows one to search by the type of object (codex or folia), material (various types of paper or palm leaf), number of leaves, height and width, collation (single or paired leaves), and certain keywords describing the condition, binding, binding material, and seal. Finally, the *administrative* pane allows one to search by the entry editor and the date of last revision.

The dynamic catalogue software generates the index from an XML driver file (*index.xml*) that describes the categories to be shown and how they are grouped in various panes, and lists the paths to the XML elements that

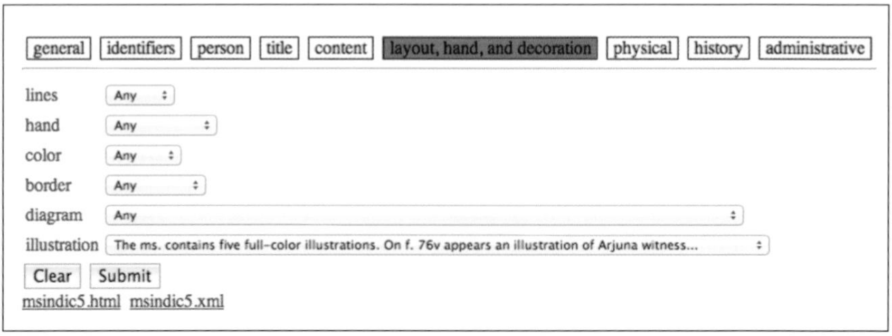

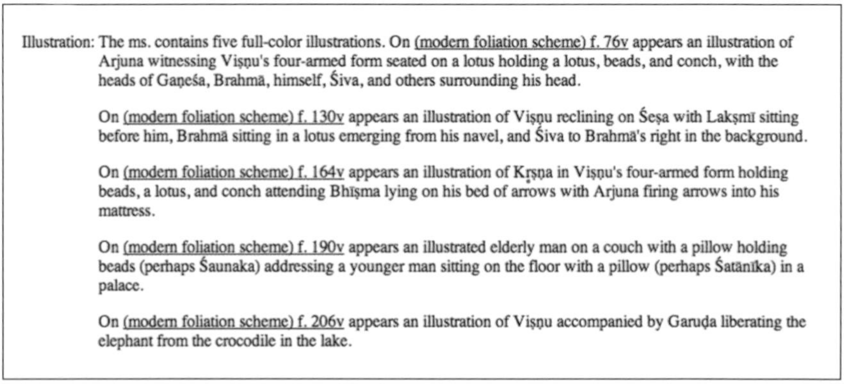

Illustration: The ms. contains five full-color illustrations. On (modern foliation scheme) f. 76v appears an illustration of Arjuna witnessing Viṣṇu's four-armed form seated on a lotus holding a lotus, beads, and conch, with the heads of Gaṇeśa, Brahmā, himself, Śiva, and others surrounding his head.

On (modern foliation scheme) f. 130v appears an illustration of Viṣṇu reclining on Śeṣa with Lakṣmī sitting before him, Brahmā sitting in a lotus emerging from his navel, and Śiva to Brahmā's right in the background.

On (modern foliation scheme) f. 164v appears an illustration of Kṛṣṇa in Viṣṇu's four-armed form holding beads, a lotus, and conch attending Bhīṣma lying on his bed of arrows with Arjuna firing arrows into his mattress.

On (modern foliation scheme) f. 190v appears an illustrated elderly man on a couch with a pillow holding beads (perhaps Śaunaka) addressing a younger man sitting on the floor with a pillow (perhaps Śatānīka) in a palace.

On (modern foliation scheme) f. 206v appears an illustration of Viṣṇu accompanied by Garuḍa liberating the elephant from the crocodile in the lake.

Figure 10.6. The layout, hand, and decoration pane in the Sanskrit Library manuscript catalogue index.

Figure 10.7. The description of illustrations in the decoration section in the Sanskrit Library manuscript catalogue HTML file for UPenn 490.

Figure 10.8. The illustration accompanying the narration of the *Gajendramokṣaṇa,* of Viṣṇu accompanied by Garuḍa liberating the elephant from the crocodile in the lake. The painting appears on fol. 206v of UPenn Ms. Indic 5.

contain the relevant information in the Sanskrit Library XML manuscript catalogue entry files. The catalogue index can thus be revised or restructured without reprogramming just by editing the XML driver file.

Text-Image Alignment

The Sanskrit Library developed the Sanskrit image-text alignment interface (SITA) to facilitate human alignment of digital images of manuscript pages with the corresponding digital text. Search and display software uses the alignment to provide dynamic direct access to individual manuscript pages that contain passages specifically sought. The context that contains the sought passage, which was aligned previously with digital text using the SITA software, is shown demarcated in each manuscript page image. Facilities to search for digital text are thus extended to digital manuscript images. This extension allows generalized information extraction and search techniques to reach Sanskrit manuscripts. The text-image alignment also allows a scholar viewing a manuscript immediate access to the digital analytic tools developed by the Sanskrit Library and its partners in the International Sanskrit Computational Linguistics Consortium, such as a parser, morphological analyzer, and digital dictionaries.

In the course of the three-year NEH-funded manuscript project, an assistant familiar with Devanāgarī and Telugu scripts and experienced in working with manuscripts used the SITA software to align some 25,000 manuscript pages with their corresponding digital texts. The SITA software displays an image of one or two manuscript pages on the left, the corresponding digital text on the right, and a comment box on the lower right. Figure 10.9 shows an example from UPenn Ms. Coll. 390, Item 490, a paper manuscript of the *Bhīṣmastavarāja*. At the left is an image of the verso of folio 1 and the recto of folio 2 of the manuscript. A section of text on the recto of folio 1 marked at its beginning and end by small red brackets is correlated with a selection of digital text highlighted in gray on the right. The selection (MBh. 12.47.1–3) is the opening of the *Bhīṣmastavarāja* (the praise of Viṣṇu by Bhīṣma), which occurs in the twelfth book (*parvan*) of the *Mahābhārata*. In the comment box at the lower right, the page on which the annotated passage occurs is written. The text in red in the manuscript image that precedes the open bracket is a passage not included in the digital edition. The passage is identified as a benediction in the comment box in another annotation while no text is selected in the digital edition.

Figure 10.10 shows an XML file of the SITA text-image alignment and *annotation* records. The first and last annotation elements record the annotations described in the previous paragraph. The attributes *x1*, *y1* and *x2*, *y2* record the coordinates of the marks in the image. The attributes *start* and *end* in the first annotation element record the beginning and end of the selection in the digital text. The bracketed *CDATA* contains the comments recording the page number "f. 1v" in the first annotation and the comment "benediction" in the last. The comment "benediction" is one item in a list of standard comments made available to the annotator and expanded as warranted in the course of the project. These comments also identify rubrics, final rubrics, colophons, text written at unusual angles, missing or additional passages, marginal additions, corrections, and bindings. The list of standard comments is shown in Table 10.1.

The page number in annotations supplies a check against the automated alignment produced from the *foliation* element of the manuscript catalogue entry in the page reference validation interface produced by the Folio software shown in Figure 10.11. Highlighted in red are the file name of the original image, the page numbers predicted by the foliation formula, and the page numbers indicated in SITA annotations. Miniatures of the origi-

---

Figure 10.9. Annotation of an image of fols. 1v–2r of UPenn Ms. Coll. 390, Item 490 in the Sanskrit image-text alignment software (SITA). The passage demarcated in red brackets on fol. 1v is aligned with *MBh.* 12.47.1–3, the beginning of the *Bhīṣmastavarāja* in the *Śāntiparvan* of the *Mahābhārata*, and the folio reference is provided in the comment box at the lower right.

Figure 10.10. XML data file containing annotations of fols. 1v–2r of UPenn Ms. Coll. 390, Item 490, produced with the Sanskrit image-text alignment software (SITA). The *annotation* elements may contain attributes *x1*, *y1* and *x2*, *y2* indicating the coordinates of the open and closed brackets demarcating a passage in the image, attributes *start* and *end* indicating the beginning and end of the corresponding passage in the digital text, and *CDATA* containing a comment.

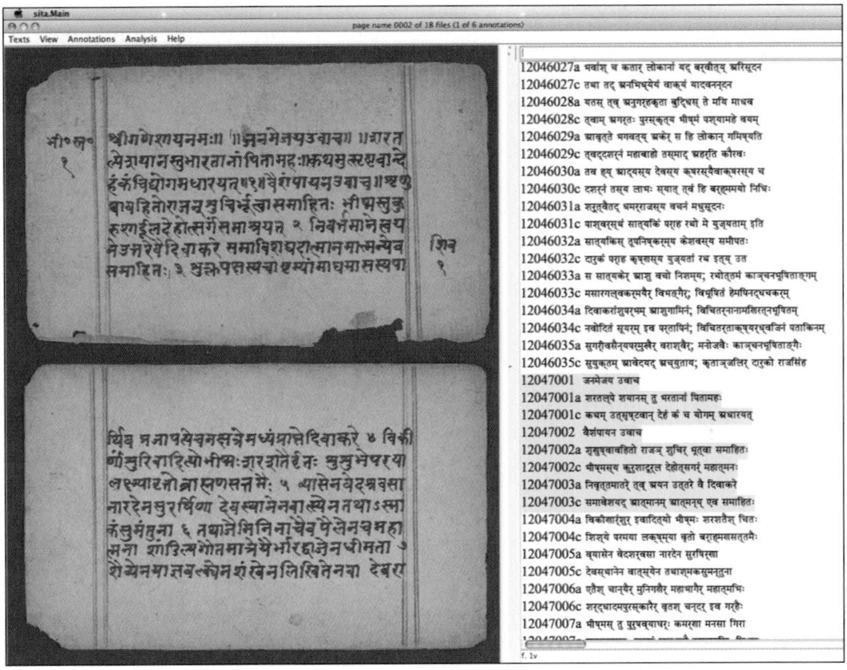


```
<annotations page='MBh12ShantiParvan/UPenn/mscoll390_item490' width='890'
    pc='-65536'>¬
Δ   <annotation type='sf' x1='1298' y1='465' x2='921' y2='1290' start='163010'
        end='163355' ><![CDATA[f. 1v]]></annotation>¬
Δ   <annotation type='sf' x1='2128' y1='2115' x2='1062' y2='2733'
        start='163356' end='163556' ><![CDATA[f. 2r]]></annotation>¬
Δ   <annotation type='sf' x1='946' y1='1195' x2='2298' y2='1265'
        ><![CDATA[passage is in manuscript but not in digital
        text]]></annotation>¬
Δ   <annotation type='sf' x1='564' y1='2120' x2='2111' y2='2223'
        ><![CDATA[passage is in manuscript but not in digital
        text]]></annotation>¬
Δ   <annotation type='sf' x1='1078' y1='2630' x2='2285' y2='2945'
        ><![CDATA[passage is in manuscript but not in digital
        text]]></annotation>¬
Δ   <annotation type='sf' x1='568' y1='465' x2='1240' y2='560'
        ><![CDATA[benediction]]></annotation>¬
</annotations>¬
```

nal image and the split and cropped images are displayed at the right. The interface allows a reviewer to check the correspondence quickly. Here the page numbers correspond, but the cropped images are in reverse order. The discovery of randomly ordered cropped images led to revision of the splitting software to constrain zone ordering based upon the zones' y-coordinates.

Table 10.1. Sanskrit image-text alignment standard comments

Comment
Ask Dr. Scharf about this
rubric
final rubric
colophon
benediction
this is commentary
text is written vertically up the right margin
text is written vertically down the right margin
text is written vertically up the left margin
text is written vertically down the left margin
passage is in manuscript but not in digital text
passage is in digital text but not in manuscript
marginal addition in the same hand to be inserted at insertion point
marginal addition in a different hand to be inserted at insertion point
marginal correction in the same hand
marginal correction in a different hand
deletion of text painted over in yellow pigment
missing text indicated by an ellipsis mark
part of the folio containing text is missing
one blank page
two blank pages
variant
misordered image file
misordered manuscript leaf
inverted manuscript image
front cover
inside front cover
back cover
inside back cover

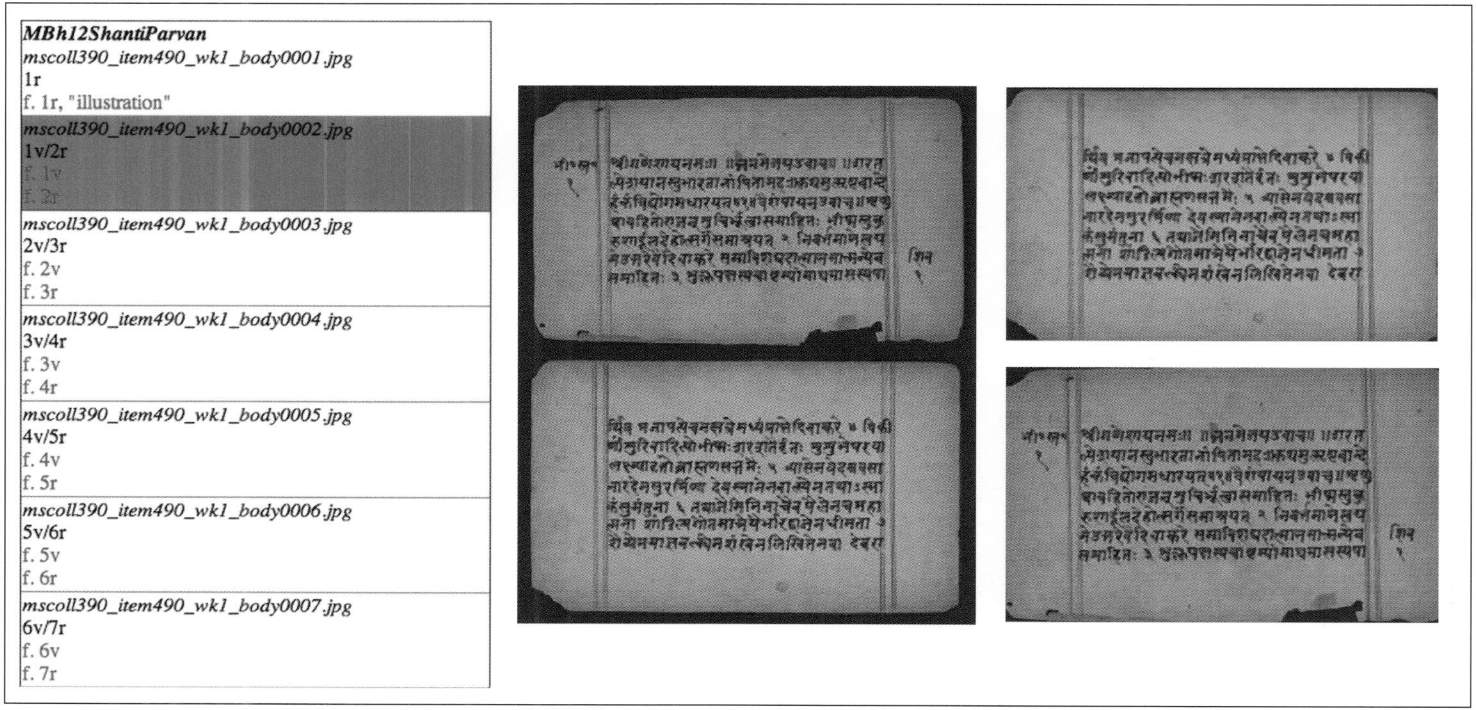

Figure 10.11. The Sanskrit Library Folio interface used to validate references to manuscript pages in image files. Highlighted in red at the left are the file name of the original image, pages predicted by the catalogue entry's foliation formula, and page numbers inserted in the comment box of the SITA software. The original image, and split and cropped images are shown at the right, the latter erroneously in inverse order here.

Integrated Digital Access to Inherited Knowledge

The XML annotation records created in SITA allow multiple manuscript images to be linked to searchable digital texts, thereby allowing focused access to individual pages in manuscript images that contain the sought passage. The text-image alignment annotations created in SITA and written to XML files were used to create the Sanskrit image-text alignment display interface that links digital text passages to the digital images showing corresponding passages demarcated with red angle brackets in each manuscript. The alignment display interface displays the digital text. While the current display is in standard Romanization, it can be displayed in the other scripts included in the Sanskrit Library's transcoding software described above (**Overcoming Obstacles to Access of Indian Heritage**). A menu in the top left corner of the interface lists all the manuscripts that contain text corresponding to the digital text. The text the selected manuscript contains is highlighted in yellow, while text it does not contain appears in white. Clicking a yellow passage in the digital text displays the manuscript image that contains the corresponding passage with the passage demarcated with red angle brackets. The passage in the digital text is colored green, and additional text contained in other annotations in the image appears in blue. Clicking the blue shows the manuscript image with the clicked passage demarcated.

The Sanskrit image-text alignment display interface is accessible from the catalogue entries of particular manuscripts as well as from the Sanskrit Library's text catalogue. Access by clicking a bibliographic scope reference in a catalogue entry was described above (**Dynamic Cataloguing: The Sanskrit Library Manuscript Catalogue HTML Display**). Clicking the scope reference *12.47.1* at the end of the incipit in the manuscript contents description of UPenn 490 shown in Figure 10.1, or at the end of verse 1 in the verse structure display in the *transcription* section shown in Figure 10.3 opens the alignment display interface shown in Figure 10.12. The HTML interface contains the text of the digital edition of the *Śāntiparvan* of the

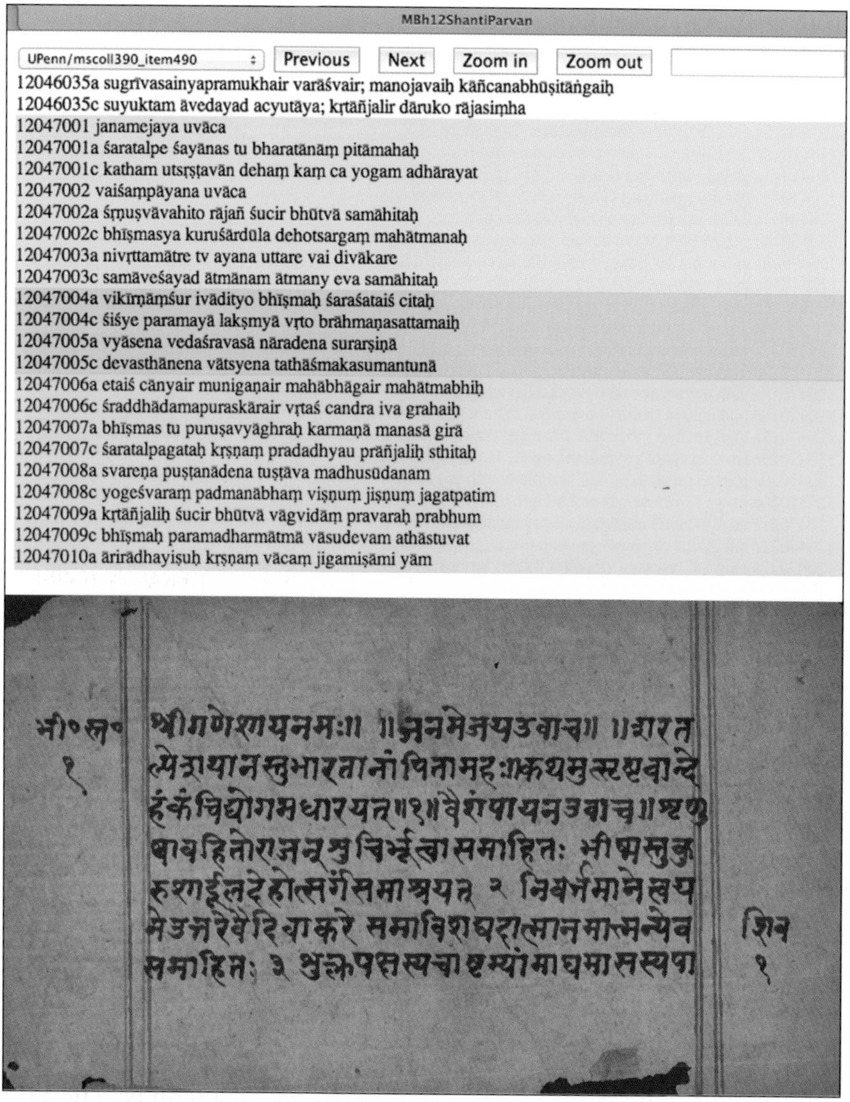
MBh12ShantiParvan
UPenn/mscoll390_item490
Previous
Next
Zoom in
Zoom out
12046035a sugrīvasainyapramukhair varāśvair; manojavaiḥ kāñcanabhūṣitāṅgaiḥ
12046035c suyuktam āvedayad acyutāya; kṛtāñjalir dāruko rājasiṃha
12047001 janamejaya uvāca
12047001a śaratalpe śayānas tu bharatānāṃ pitāmahaḥ
12047001c katham utsṛṣṭavān deham kaṃ ca yogam adhārayat
12047002 vaiśampāyana uvāca
12047002a śṛṇuṣvāvahito rājañ śucir bhūtvā samāhitaḥ
12047002c bhīṣmasya kuruśārdūla dehotsargaṃ mahātmanaḥ
12047003a nivṛttamātre tv ayana uttare vai divākare
12047003c samāveśayad ātmānam ātmany eva samāhitaḥ
12047004a vikīrṇāṃśur ivāditya bhīṣmaḥ śaraśataiś citaḥ
12047004c śiṣye paramayā lakṣmyā vṛto brāhmaṇasattamaiḥ
12047005a vyāsena vedaśravasā nāradena surarṣiṇā
12047005c devasthānena vātsyena tathāśmakasumantunā
12047006a etaiś cānyair munigaṇair mahābhāgair mahātmabhiḥ
12047006c śraddhādamapuraskārair vṛtaś candra iva grahaiḥ
12047007a bhīṣmas tu puruṣavyāghraḥ karmaṇā manasā girā
12047007c śaratalpagataḥ kṛṣṇaṃ pradadhyau prāñjaliḥ sthitaḥ
12047008a svareṇa puṣṭanādena tuṣṭāva madhusūdanam
12047008c yogeśvaraṃ padmanābhaṃ viṣṇuṃ jiṣṇuṃ jagatpatim
12047009a kṛtāñjaliḥ śucir bhūtvā vāgvidāṃ pravaraḥ prabhum
12047009c bhīṣmaḥ paramadharmātmā vāsudevam athāstuvat
12047010a ārirādhayiṣuḥ kṛṣṇaṃ vācaṃ jigamiṣāmi yām

Figure 10.12. The Sanskrit image-text alignment display interface (SITADI). The selected passage shown in green, *MBh.* 12.47.1–3, the beginning of the *Bhīṣmastavarāja* in the *Śāntiparvan* of the *Mahābhārata*, is a link that displays the passage demarcated with red brackets in the manuscript selected in the menu at the top left. Other text annotated in the displayed image appears in blue, and other text annotated in the manuscript appears in yellow.

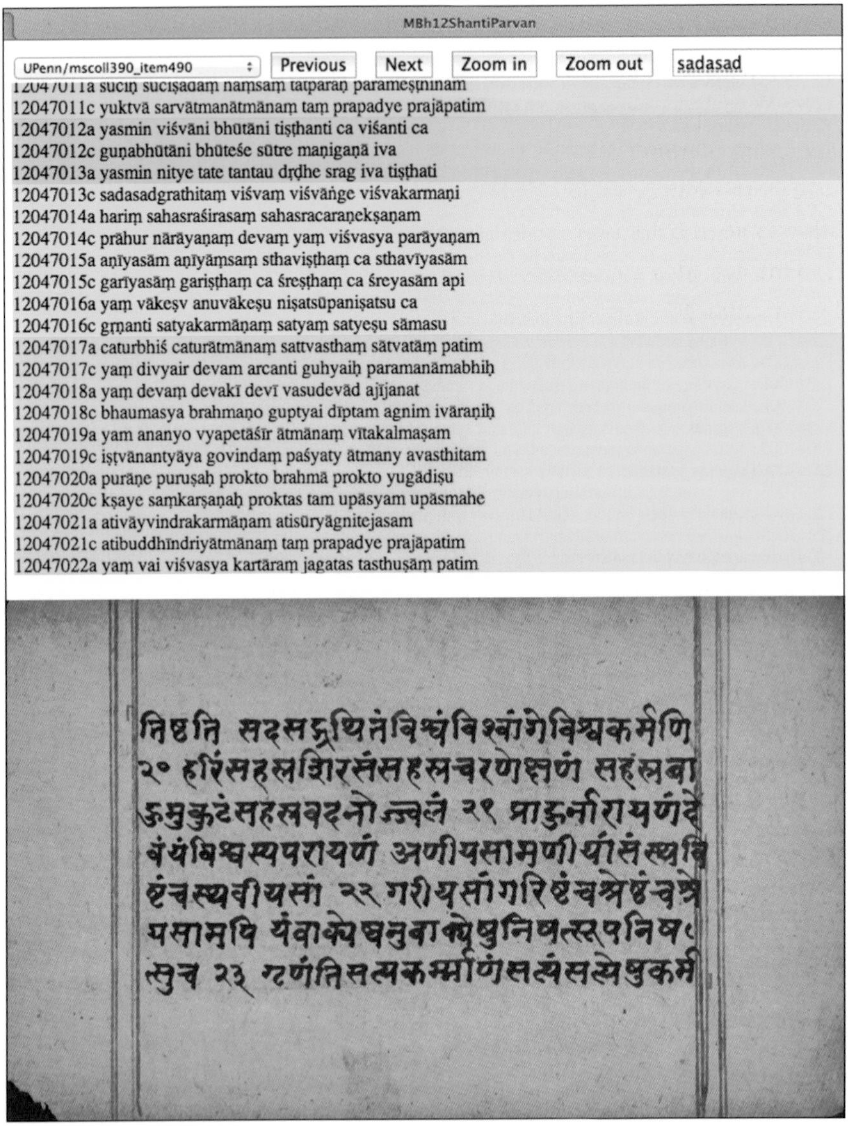

Figure 10.13. Image showing the Sanskrit image-text alignment display interface (SITADI). The sought string *sadasad* was located in the passage shown in green, *MBh.* 12.47.13c–16c, in the manuscript shown in the menu at the top left, that is, UPenn Ms. Coll. 390, Item 490. The page of this manuscript linked to the passage in green is shown below.

Mahābhārata scrolled to the beginning of the *Bhīṣmastavarāja*. The incipit of the text is shown in green, and the manuscript image with the corresponding passage demarcated is displayed below it. By selecting other manuscripts from the menu in the upper left corner, the corresponding passage in those manuscripts is displayed. A search box at the top of the page allows one to search for other passages in that text.

One may also access the Sanskrit image-text alignment display interface by searching for a passage in the Sanskrit Library text window. For example, searching for *sadasad* ("existent and non-existent") opens the alignment display interface with the passage located in the first manuscript in which it is found. Selecting UPenn 490 from the menu displays the passage in that manuscript as shown in Figure 10.13.

The comments in annotations made in SITA allow one to locate annotations of a particular type or annotations whose comments contain a particular term. Searching for these comments provides access to a number of interesting features of manuscripts. An interface will soon be built to provide access to manuscript images via the list of standard comments shown in Table 10.1 or by search of any text in comments. At present, they are searchable by a text editor.

Sharing Expertise for the Propagation of Knowledge and Culture

Digital technology is flexible. This flexibility allows incremental development of software as well as progressive addition of data. The Sanskrit Library manuscript catalogue's extensive manuscript description apparatus may seem overwhelming, yet the flexibility of digital technology permits data to be accumulated in whatever state it may presently be available. Hence, it would be easy to add data available in simpler manuscript lists regardless of how large or small and regardless of how detailed. Small institutions and private collections are often overlooked by scholars in their search for manuscripts. The Sanskrit Library is therefore particularly eager to provide centralized access to such minor and private collections. The procedure by which one would produce a hypertext catalogue and searchable access to manuscript images is described in the author's article, "Accessing Manuscripts in the Digital Age: A Pipeline to Create a Hypertext Catalogue, and Searchable Access to Manuscript Images," in the Tattvabodha series published by the Indira Gandhi National Centre for the Arts in New Delhi.

The Sanskrit Library developed software to convert its TEI manuscript catalogue to the machine-readable catalogue format used by most libraries in the United States and United Kingdom (MaRC) and makes the catalogue entries that describe manuscripts available in this form to the libraries that house those manuscripts so that they may be included in standard catalogues at those institutions.

The protocols, formats, and software described above are compatible with Web-accessible digital images of Sanskrit manuscripts wherever they are hosted. The SITA software is already being used in a project in Kerala, India, and funneling of a new entry through the cataloguing pipeline was recently demonstrated at a workshop in Maharashtra. The Sanskrit Library is currently engaged in cataloguing the entire collection of seventeen hundred Sanskrit manuscripts in the Houghton Library at Harvard University and intends to pursue projects to catalogue and digitize manuscripts at collections of Sanskrit manuscripts nationwide in the United States. We are likewise eager to collaborate with institutions in India and worldwide to catalogue and digitize manuscripts wherever they may be found.

Appendix A. Sanskrit Library Manuscript Cataloguing Template

The following is a description of the Sanskrit Library's manuscript cataloguing template structured from elements described in the Text Encoding Initiative's (TEI) Manuscript Description guidelines (www.tei-c.org/release/doc/tei-p5-doc/en/html/MS.html). The TEI Manuscript Description guidelines allow for the complete transcription of a manuscript in the body of a TEI document; the TEI header element (*teiHeader*) contains three subsections to describe the file, its encoding, and its categorization or profile:

1. fileDesc
2. encodingDesc
3. profileDesc

The *fileDesc* element includes an element for the description of the document source (*sourceDesc*), which in turn includes an element for the description of the manuscript (*msDesc*). This *msDesc* element provides elements to describe the manuscript that serve as a thorough catalogue entry. In the following description, numerals and letters in parentheses refer to the outline of those subelements in the table below. Nested element names are separated from the name of their parent by a dot, and attributes are put in square brackets. The template includes categories for the description of manuscript identifiers, manuscript contents, the physical condition of the manuscript, its history, and any additional information (main headings I–V).

Included in the TEI manuscript header's *msIdentifier* element are elements to designate the collection in which the manuscript is housed and its identifying number in that collection (I.A–B). Provision is also made for describing alternate identifiers that may describe the identification of the manuscript in catalogues (I.C–D).

Included in the TEI manuscript header's *msContents* element is an *msItem* or *msItemStruct* element that describes the content of a work or a part of a work (II.A). The element may be repeated for each work contained in a manuscript and may be nested to describe sections or subsections of a work. The *msItem* and *msItemStruct* elements contain an indication of whether the manuscript is complete or not and elements to describe the author, title, headings (*rubric*), beginning of the work proper (*incipit*), end of the work proper (*explicit*), its closing (here called "finalRubric") scribal trailer (here called "colophon"), and its language and script (II.A.1–9). The *note* element includes elements to mark names of people, titles, bibliographic information, and so forth (II.A.8.a–c). While the *msItem* element allows freer structuring and repetition of elements, the *msItemStruct* preserves information included in the following order for simpler automated processing.

The *physDesc* element (III) includes elements to describe the physical aspects of the manuscript, including the type of object (folia or codex), its material basis (*support\-Desc*), arrangement of the text on each page (*layoutDesc*), a description of the scribal hands (*handDesc*), decorations (*decoDesc*), interlinear and marginal notes and corrections (*additions*), the binding (*bindingDesc*), the seal (*sealDesc*), and any accompanying matter such as letters of donation or acquisition forms (*accMat*) (III.A–G).

Description of the material basis (*supportDesc* III.A.1) includes elements that describe the material (paper or palm leaf), size and quantity of leaves, their enumeration (*foliation*), and their collation (these latter two include regular formulas amenable to digital processing), abbreviated titles that accompany enumeration (*signatures*), and the condition of the material basis (III.A.1.a–i). Description of the layout (*layoutDesc* III.A.2) indicates how many columns and lines are on each page and anything else that may describe the layout of the text. For example, the *layout* element may describe that a base text is indented, flanked above and below by commentary in an hourglass arrangement.

The *handDesc* (III.B) element includes an element to summarize the scribal hands, if necessary, as well as a repeatable element to describe the handwriting of each scribe (III.B.1–2). Description of the decoration (III.C) is done by the use of *decoNote* elements provided with various values of a type attribute to distinguish description of colors, borders, illustrations, and diagrams (III.C.1–4). The *additions* element (III.D) allows one to describe interlinear and marginal notes. The *bindingDesc* element (III.E) lets one indicate whether there is a binding, what sort of binding it is, the materials it uses, and its condition. The *accMat* element describes letters, acquisition documents, wrappers, and other materials that may accompany the manuscript (III.G).

The TEI manuscript header's *history* element includes elements to describe the origin of the manuscript as well as facts about its subsequent locations, ownership, and use (provenance) (IV.A–C). Elements are provided to mark the date when it was completed (*origDate*); the names of the scribe, patron, and owner (*persName*); and the place of origin (*placeName*, *geogName*) (IV.A.1–4). These name elements may also be used to describe details of the manuscript's provenance (IV.A) and its acquisition by the current repository (IV.B).

The TEI manuscript file description closes with the additional element that includes a record history element (*recordHist*) subsumed under an administrative information element (*adminInfo*); these elements describe the source of the information that is included in the XML catalogue record itself and any changes made to that record (V.A–B).

The following are the principal XML elements described by the TEI Manuscript Description guidelines included in the Sanskrit Library's manuscript catalogue template:

I. msIdentifier
    A. collection
    B. idno
    C. altIdentifier[type='catalog'].collection
    D. altIdentifier[type='catalog'].idno
II. msContents
    A. msItemStruct[defective='true|false'
        1. rubric
        2. author
        3. title
        4. incipit
        5. explicit
        6. finalRubric
        7. colophon
        8. note
            a. persName
            b. title
            c. bibl
        9. textLang[mainLang='ll-Ssss']

III. physDesc
    A. objectDesc[form='folia|codex']
        1. supportDesc
            a. support.material
            b. extent.measure
            c. extent.dimensions
            e. foliation
            f. foliation.formula
            g. foliation.signatures
            h. collation
            i. collation.formula
            j. condition
        2. layoutDesc
            a. layout
            b. layout[columns]
            c. layout[writtenLines]
    B. handDesc
        1. summary
        2. handNote
    C. decoDesc
        1. decoNote[type='color']
        2. decoNote[type='border']
        3. decoNote[type='illustration']
        4. decoNote[type='diagram']
    D. additions
    E. bindingDesc
        1. binding
        2. condition
    F. sealDesc.seal
    G. accMat
IV. history
    A. origin
        1. origDate
        2. persName type='scribe|owner'

        3. placeName
        4. geogName
    B. provenance
    C. acquisition
V. additional.adminInfo.recordHist
    A. source
    B. change

Appendix B. Sanskrit Library Indic Subject Classification

I. Śruti. Aural knowledge, in particular Veda
    A. Saṁhitā. Continuous unalterable text
        1. Ṛgveda. Collection of verses (*mantra*)
        2. Sāmaveda. Collection of verses recited in seven tones
        3. Yajurveda. Collection of verses and prose
        4. Atharvaveda. Collection of verses
    B. Brāhmaṇa_main. Vedic prose
        1. Brāhmaṇa. Prose commentary on ritual and text used in it
        2. Āraṇyaka. Forest books
        3. Upaniṣad. Private instruction
II. Smṛti. Remembered or traditional knowledge
    A. Vedāṅga. Limbs of the Veda
        1. Śikṣā_main. Phonetics
            a. Prātiśākhya. Phonetic texts of particular ancient Vedic schools
            b. Śikṣā. Later phonetic treatises
        2. Kalpa. Ritual
            a. Śrautasūtra. Public ritual
            b. Gṛhyasūtra. Domestic ritual
        3. Vyākaraṇa. Grammar
            a. Pāṇinīya. Pāṇinian Grammar
            b. Apāṇinīya. Non-Pāṇinian Grammar

4. Nirukta. Etymology of Yāska, commentator on Vedic word lists (*nighaṇṭu*)
5. Chandas. Metrics, Music
6. Jyotiṣa. Astronomy, Astrology

B. Darśana. Subordinate limbs of veda (*upāṅga*) or philosophical views

1. Nyāya. Epistemology, Logic, and Argumentation founded on Gautama's *Nyāyasūtra*
2. Vaiṣeṣika. Ontology founded on Kaṇāda's *Vaiśeṣikasūtra*
3. Sāṅkhya. Evolutionary Ontology founded on Kapila's non-extant work
4. Yoga. Practice founded on the *Yogasūtra* attributed to Pata-njali
5. Karmamīmāṁsā. Ritual Exegesis founded on Jaimini's *Pūrvamīmāṁsāsūtra*s
6. Vedānta. Metaphysics founded on Bādarāyaṇa's *Uttaramīmāṁsāsūtra*s

C. Upaveda. Subordinate Veda

1. Āyurveda. Medical Science such as the works of Caraka, Suśruta, Vāgbhaṭṭa, et al. (including veterinary medicine and arboriculture)
2. Gandharvaveda_saṅgītaśāstra. Music
3. Dhanurveda. Military Science
4. Sthāpatyaveda_vāstuśāstra. Architecture and Environmental Engineering

D. Dharmaśāstra. Duty, Custom, and Law

1. Dharmasūtra. Rules of particular Vedic schools

E. Itihāsa. Narrative, Epic, History

1. Mahābhārata
2. Rāmāyaṇa

F. Purāṇa. Ancient Cosmogony, Genealogy, Narrative

III. Bhakti. Devotional literature (including stotras)

IV. Tantra. Esoteric ritualism

A. Āgama. Authoritative literature in Tantra

B. Mantra. Compendia of mantras used in Tantra

V. Kāvya. Fine literature and poetry (belles lettres)
VI. Kathā. Story literature other than Itihāsa and Purāṇa
VII. Kośa. Dictionaries and Thesauri, including nighaṇṭus
VIII. Alaṅkāraśāstra. Literary criticism
IX. Nāṭyaśāstra. Dance and Theater
X. Śilpaśāstra. Arts and Crafts
XI. Arthaśāstra. Politics, economics, statecraft (including Nīti)
XII. Ratnaśāstra. Gemology
XIII. Kāmaśāstra. The science of making love
XIV. Rasaśāstra. Alchemy
XV. Saṅgraha. Encyclopedias

Notes

1 For more detail, see Peter Scharf and Malcolm Hyman, *Linguistic Issues in Encoding Sanskrit* (Delhi: Motilal Banarsidass; Providence: Sanskrit Library, 2011).

2 Peter Scharf, "Vedic Accent: Underlying versus Surface," in *Devadattīyam: Johannes Bronkhorst Felicitation Volume*, ed. François Voegeli et al., 405–26 (Bern: Peter Lang, 2012).

3 *Sanskrit Computational Linguistics: First and Second International Symposia, Rocquencourt, France, October 2007; Providence, RI, USA, May 2008; Revised Selected and Invited Papers*, ed. Gérard Huet, Amba Kulkarni, and Peter Scharf, Lecture Notes in Artificial Intelligence 5402 (Berlin: Springer-Verlag, 2009); *Sanskrit Computational Linguistics: Third International Symposium, Hyderabad, India, January 2009, Proceedings*, ed. Gérard Huet and Amba Kulkarni, Lecture Notes in Artificial Intelligence 5406 (Berlin: Springer-Verlag, 2009); *Sanskrit Computational Linguistics: Fourth International Symposium, New Delhi, India, December 2010, Proceedings*, ed. Girish Nath Jha, Lecture Notes in Artificial Intelligence 6465 (Berlin: Springer-Verlag, 2010); *Proceedings of the Fifth International Sanskrit Computational Linguistics Symposium (4–6 January 2013, IIT Bombay, Mumbai)* (New Delhi: D. K. Printworld, 2013).

Contributors

Angela S. Chiu is a research associate in the Department of the History of Art and Archaeology at the School of Oriental and African Studies, University of London.

Alexandra Green is the Henry Ginsburg Curator for Southeast Asia at the British Museum in London. Her publications include *Burma: Art and Archaeology* (2002), coedited with Richard Blurton; *Eclectic Collecting: Art from Burma in the Denison Museum* (2008); and *Rethinking Visual Narratives from Asia: Intercultural and Comparative Perspectives* (2013).

Justin Thomas McDaniel is a professor of Buddhist studies and chair of the Department of Religious Studies at the University of Pennsylvania. He is the author of *The Lovelorn Ghost and the Magic Monk: Practicing Buddhism in Modern Thailand* (2011) and *Gathering Leaves and Lifting Words: Histories of Monastic Education in Laos and Thailand* (2008), winner of the Harry Benda Prize for Best First Book in Southeast Asian Studies.

Kim Plofker is assistant professor of mathematics at Union College in New York. She is the author of *Mathematics in India* (2009) and many articles on the history of the exact sciences in India, Islam, and early modern Europe.

Lynn Ransom is the project manager of the Schoenberg Database of Manuscripts and a founding member of the Schoenberg Institute for Manuscript Studies at the University of Pennsylvania Libraries.

Peter M. Scharf has held visiting professorships at the Maharishi University of Management Research Institute, the Indian Institute of Technology Bombay, the University of Hyderabad, and the University of Paris Diderot. His research focuses on the linguistic traditions of India, Vedic

Sanskrit, and Indian philosophy. He is the founder and director of the Sanskrit Library (sanskritlibrary.org).

Daniel Sou is a lecturer at Pennsylvania State University, Abington, where he teaches East Asian history and religion. He has published in *Monumenta Serica* on the fundamental disciplines for a Qin local official recorded in excavated texts and is pursuing research on how early Chinese empires established, expanded, and solidified central authority through legal communication.

Ori Tavor teaches Chinese history, religion, and thought in the Department of East Asian Languages and Civilizations at the University of Pennsylvania. His research examines the impact of individual biospiritual practices on the development of ritual theory in premodern China and the role of technical manuscripts in the production of early Chinese discourses on the body.

Sergei Tourkin is an independent scholar. He has held positions at the Institute of Oriental Studies in St. Petersburg, Russia; McGill University in Montreal; and *Encyclopaedia Iranica* at Columbia University.

Sinéad Ward is a doctoral student at the School of Oriental and African Studies in London, where her research focuses on the decorative *Kammavācā* manuscripts of Burma. She is also rights and reproductions officer at the Chester Beatty Library, Dublin.

Susan Whitfield is curator of the Dunhuang and Central Asian Manuscript Collections at the British Library and head of the International Dunhuang Project. She has published widely on the history, archaeology, and art of the Silk Road.

Hiram Woodward is the Mr. and Mrs. Thomas Quincy Scott Curator of Asian Art, Emeritus, at the Walters Art Museum in Baltimore, Maryland. He is the author of *The Art and Architecture of Thailand: From Prehistoric Times Through the Thirteenth Century* (2003).

Index